Piece of Another Period

Four sons and a Cambridge childhood

By

Richard, Stephen, Robert and John Boyd

**Grosvenor House
Publishing Limited**

The right of Richard, Stephen, Robert and John Boyd
to be identified as the authors of this
work has been asserted in accordance with Section 78
of the Copyright, Designs and Patents Act 1988

This book is published by
Grosvenor House Publishing Ltd
Link House
140 The Broadway, Tolworth, Surrey, KT6 7HT.
www.grosvenorhousepublishing.co.uk

A CIP record for this book
is available from the British Library

Paperback ISBN 978-1-83615-163-0
eBook ISBN 978-1-83615-164-7

It really is Cambridge, it really is ... I feel so very provincial coming from the medical school of Belfast.
[Dixon to Amélie, 31 July 1932]

Keep enclosed letters for me – will you; they might be nice for our grandchildren.
[Amélie to Dixon, 27 June 1933]

I do hope your letters to me will survive the vagaries of the war as I think they will give the boys pleasure sometime in the distant future.
[Dixon to Amélie, 24 April 1941]

Contents

The Family   ix

Introduction   xi

1 Before Cambridge   1
i   Young Dixon and Amélie   1
ii   Medical Students   6
iii   Hugh Meredith – A Cambridge Apostle in Belfast   9
iv   League of Nations Summer School   14
v   Graduation and the Road to Marriage   18
vi   Two Surgical Cousins   28
vii   Cambridge Job Offer   32
viii A Year in America   33
ix   Amélie and Dixon as Medical Practitioners   35

2 Cambridge on the Margins of War   47
i   Getting Established in Cambridge   47
ii   London Briefly   51
iii   Separated by the Irish Sea   52
iv   Wartime Cambridge   67
v   Anatomists as Family   69
vi   Back in London   75

3 A Cambridge Family   81
i   Appointment   81
ii   Visitors   84
iii   Three Aunts   86
iv   A Female Scientist   95
v   Another Belfast Anatomist   96

4 Childhoods   99
i   Schools   99
ii   Transport and the River   100
iii   Reading   106
iv   Other Mothers and Fathers   109

v	Hobbies and Pets	115
vi	Religion and Mental Health	118
vii	Boys and Girls (and Sex)	124
viii	Sons of the Professor	128
5	**Family Life**	**135**
i	Help and Childcare	135
ii	Politics	145
iii	Jews, Blacks, and Irish	147
iv	Money	153
6	**Who Knew Who**	**157**
i	Scientists	157
ii	Darwins and Keyneses	165
iii	The Adrians	169
iv	AV Hill	171
7	**Anatomy and Academia**	**173**
i	Research	173
ii	Professorial Politics	180
iii	Teaching	184
iv	Scholarship	185
v	Clare College	188
8	**Amélie and Dixon**	**195**
i	A Cambridge Wife	195
ii	Dixon and his Ashtray	204
9	**Cycle of Life**	**207**
Index		**211**
Acknowledgements		**233**
Related volumes		**235**
The Authors		**237**

Illustrations

1. Some of the Letters   xiii
2. First Name Terms?   7
3. House Physician   19
4. Coffee at the Carnegie Institute Baltimore   35
5. Parenthood   48
6. Wartime   57
7. Boyd Boys 1   80
8. Moving House   156
9. The Boyds at 21 Newton Road Cambridge   156
10. Ray Club Dinner 25 May 1957   159
11. Cambridge Graduate Sciences Club Programme 1954-5   160
12. A Christmas Dinner at Clare   161
13. Daughters of Cambridge Dynasties   167
14. Dixon and Microscope   175
15. In the Womb   177
16. Fellows, and some Wives, processing over Clare Bridge   189
17. Boyd Boys 2   207

Reminiscences

1. HOM and Richard   11
2. James Bartley   12
3. At the Townsleys   30
4. Elizabeth and Richard   85
5. Richard's Bicycle   101
6. Richard's Gang   104
7. Early Morning Swim   105
8. Richard's Divinity Class   118
9. Robert's Extra-curricular Education   125
10. An Oxford Professor   129
11. Old Mrs Keynes   167
12. Adrian Keynes   170
13. Richard as Chauffeur   178
14. Walter Bodmer – the 'New Science'   180

Note on quotations: Almost all longer quotes, and most short ones, are from letters or other documents; a few are memories of one of the authors. Spelling and punctuation are generally as written but have been occasionally modified for ease of reading. (Transcripts of originals are available from r.boyd@manchester. ac.uk). Inserts to explain or clarify are in [brackets]. Illegible passages are marked [?]. <u>Underlining</u> is as in original.

Cover: Front Cover. Rosie Scott - Clare College from Clare Bridge
Back Cover. Fellows on Clare Bridge (see illustration 16)

The Family

Professor of Anatomy: Dixon Boyd (1907-68) – father (Daddy)
        Grace Smyth – his mother
        James Dixon Boyd (Sr) – his father
        Norman and Gerald Townsley – surgeon cousins

Cambridge wife: Dr Amélie Loewenthal then Boyd (1906-98) – mother (Mummy, Ma)
        John McCaldin Loewenthal (JMcC) – her father
        Elsa Iklé – her mother
        Helen, Joan, Peggy – her sisters
        Iklé relatives – numerous
        Eddie Bennet – psychiatrist cousin

Their boys:
        John (1936-2019)
        Robert (1938-)
        Stephen (1943-)
        Richard (1945-)

Introduction

On 19 August 1958, Dixon Boyd, Cambridge professor, wrote to his wife, Amélie. She was camping in Scotland with their two younger sons. It was the day before their 25th wedding anniversary.

> *21 Newton Rd Cambridge*
> *Darlingest,*
> *It not only seems but it is a long time ago. We were half our present ages; the war was no more than a shadow of a man's hand in bright sunlight – possible but not very probable; there was no John or Robert or Stephen or Richard. There were your father, my father … In a wider field no atomic power or bombs; no television; unemployment and a servant class; no penicillin; no Russian imperialism and a well-established British Empire. We had not even dreamed of Cambridge, let alone three periods of residence and a Chair and Clare.*

This is the story of Cambridge in those mid-century years as it was lived by one couple, Dixon and Amélie, and experienced by their four sons: John, Robert, Stephen, and Richard, 'Three periods of residence and a Chair and Clare'. Clare was Clare College.

Dixon died within a decade of writing the letter; smokers died young in those days. Widowed, Amélie stayed on in Newton Road. Only when she eventually left Cambridge for Oxford to live next door to Richard did she empty the papers from her polished, bow-fronted desk into a black suitcase which she deposited in Robert's garage.

In time, the metal fittings of the suitcase rusted. It was old and bore the initials of Elsa, Amélie's mother, our grandmother, but only an absent one to us. More time, and the contents of the suitcase migrated to a box alongside other detritus of school reports and student memorabilia left behind in turn by one of her fourteen, now middle-aged, grandchildren. That box, tipped out in 2019, delivered hundreds of letters. Not just those between Amélie and Dixon from medical student times to their last years together, but also from those who influenced their lives or whom they influenced: relatives and friends; servants and vice-chancellors; minor colleagues and, occasionally, scientific superstars. The couple wrote whenever apart, sometimes daily. The need for urgent communication only occasionally led to what was then a *wire* or a *trunk call*. Of the children, John, and Stephen – boarding-school-educated and later living abroad – are vigorous letter-writers and recipients of many parental letters. Robert and Richard, who never went away to school, received fewer. Those two, unlike John and Stephen, followed the parents into medicine and science; memory of shared gossip in those domains sometimes provides information absent from the correspondence.

Inspired by reading the suitcase-letters, Richard searched his loft and found a second box, the letters Amélie had taken with her to Oxford. Piecing together this double archive has been great fun; spiced with smiles of recognition as well as occasional tears. John, the eldest of us, died before all the letters were uncovered but, always a doughty sentimentalist, contributed fully to the moisture, as well as to the laughter. Stuck in London in his final months he thought there might be something at his family farm. At the farm, during the 2020 Covid lockdown, his daughter, Jessica, Amélie's granddaughter, searched, and found more than something, another large cache of letters. Stephen, combing through his house in Osaka, then found more.

In '50s Cambridge, if one of us four *Boyd Boys* pedalled, from our then-home in Grange Road, over the river, ignoring the occasional Silver Street car, to wobble onto the towpath past Scudamore's punt yard, we would sometimes notice an old woman in a brown coat. In a garden just across the water she could be seen hunched forward, quite still but apparently occupied. She was Gwen Raverat, said Amélie ('Mummy', who knew most university wives and widows) who was, said Dixon ('Daddy', who rarely missed an opportunity to increase our 'general knowledge'), Charles Darwin's granddaughter. They didn't add that she had been a neopagan with Rupert Brooke and an intimate of Virginia Woolf; perhaps they didn't know that. Raverat's husband had died long before and she filled her hours sketching. That is why she sat in the garden of her adjacent house,

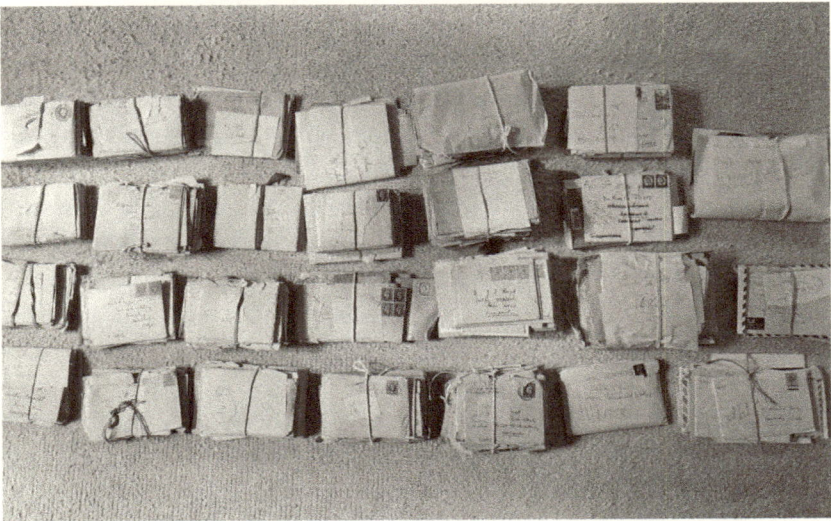

1. Some of the letters.

so apparently preoccupied, delineating the locations which her remembered Cambridge childhood inhabited.

Period Piece is her account of that childhood. Amélie adored it and used to give the volume to friends who visited. A copy lay on the table at 'her side' of the parents' double bed. It has a blue cover on which is depicted the author's memory of that Silver Street Bridge when she was young. The work, Richard points out, sets a dauntingly high bar for any attempted account of children's lives in a Cambridge academic family. Her chapters range widely over such topics as 'Education' or 'Ladies', or 'Society', or 'Uncles' presented as 'Spokes of the wheel from the hub, which is me'.

Raverat's parents were married 49 years before Amélie and Dixon. In both cases, one of the couple was US-born. Her Cambridge was sustained by the apogee of High Victorian imperial supremacy – college life at its grandest and most stable. That, at least, is an apogee she conveys. Only 50 years later, change is, as Dixon's letter points out, the dominant theme.

Her tone is witty and her memory given life through the people she remembers and the fruits of her sketching. Ours is a different Cambridge, the Cambridge of a more mundane family whose origin is provincial Irish, not one birthed in the Cambridge aristocracy. It is the Cambridge of a laboratory-busy medical scientist, not the stately world of Gwen's relations, and a Cambridge war-experienced and economically constrained. Our Cambridge 'period' is, inevitably, a more sombre and more practical one than hers (and one, alas, without the support of her brilliant visual sense).

Her 'hub' is herself. Our hub was, and in those years is, Amélie and Dixon. To us children, they are normality and are permanent. Our lives joined with them at breakfast and at supper. In between, Amélie gave life to the home. If she was out 'doing a clinic' when we returned from school, the house would feel empty. Dixon, by contrast, would have biked off to 'the department' after breakfast. This was the way, with minor variations, things were. We grew up and changed. They were our constant. Dixon's routine of return – bicycle into the garage, cycle-clips off, and, as he came through the door, the two-tone whistle of greeting to Amélie – an important element.

Letters provide some riposte to the regretted failure to press for an answer from those long gone *when they were there*. Scrawled in pencil on scrap-paper, typed on aerogrammes, fountain-penned over a dozen sides (no ball-points or biros then) or dashed off on a postcard they, indeed, were there. Bills, receipts and tax returns, also in the suitcase, contribute to the immediacy. So does the parental handwriting; hers more legible than his.

A parent is not a fixed point, though children may think so. The letters prompt us to view again the trajectory of Dixon and Amélie as a couple. The fixity was transient. Their evolving balance and dynamic as they mature and then age have lessons, rather late in the day you may say, for how we evaluate our lives and theirs. In reading, we have, all four, certainly found ourselves left with not just greater understanding but also a warmer appreciation, of them as parents and as humans. Cambridge – the university, not the town which was almost invisible to us – and the Boyds as part of Cambridge were also normality. But no; the life of this mid-century family and that mid-century university are now further in time from today than Raverat's childhood was from ours. The letters forcibly remind us, again late in the day, that normality, public as well as private, is indeed always transient.

In use of memory to qualify and understand a letter, our difference in age, nine years, feels trivial. In 1945 – war-veteran John and Robert or two-year-old Stephen and newborn Richard – it was profound and can still, we find, influence ongoing cultural assumptions.

Cambridge was, for this family, heralded by a letter.

31/12/34

My dear Boyd, my best wishes for a Happy and Prosperous New Year to you and Mrs Boyd. Will you be so good as to send me a sheet with

1. *Academic record*
2. *Teaching and Clinical experience*
3. *Publications*

I would like to receive this as soon as possible so that the Appointments Committee may ratify your appointment as University Demonstrator.
[H. A. Harris, then newly appointed Cambridge professor of anatomy, to Dixon]

1 Before Cambridge

i   Young Dixon and Amélie

According to Dixon, they first met in 'Hughie Graham's chemistry lectures'. Those were at Queen's University Belfast (Queen's or QUB, not to be confused with Queen's College Cambridge). It must have been in 1924, following a decade which, for Ireland, had seen the Somme, the Easter Rising and the firing squad, the Black and Tans, the Belfast riots, Partition, and the new Republic's civil war. None of these traumas are mentioned in their letters; life moves on.

Dixon's background rendered their later marriage improbable and success in academic life unlikely (though formally James Dickson, he was always known as Dixon, occasionally JDB, never James and always with an x). He had been born in New York to Protestant Irish immigrants. Grace, his mother, died of consumption, the dreaded 'TB'.

> *I go down so quickly, fade like a flower in half a day and then pick up just as readily ... I am getting used to dying and I guess by the time I am ready to cash my check I will have gotten used to it.*
> [Grace Boyd (née Smythe), Newark, New Jersey, to sister Margaret, undated, about 1912]

Her son, our father, had a tuberculous neck-scar from his early childhood with her. After the war – 'the Last War', not 'World War I' – and a failed second marriage, his father (also Dixon) gave up transatlantic aspirations and returned to County Antrim with the 13-year-old. When John and Robert knew that grandfather (not Stephen and Richard; he died the month Richard was born) he was running a little sweet shop post office in Greenisland, along the Lough from Belfast. We enjoyed cancelling, with a rubber stamp and an ink pad, the stamps on outgoing letters. Our father had lived there as a teenager with Lily, his second stepmother, and two half-brothers.

1

Amélie, one of four sisters (the 'Loewenthal Girls' – Helen, Amélie, Joan and Margaret, known as Peggy) was cut from another cloth, large-scale lace and linen businesses and social prominence. Her mother, Elsa, was an Iklé. Iklé Frères were lace-manufacturers with branches in Hamburg, St Gallen, Paris, London, and New York. Her father, John McCaldin Loewenthal, known as JMcC, inherited a Belfast linen export-firm. He was, by birth, a second cousin of Elsa. Elsa was fun and lively and active.

> *Yesterday we went to Stormont ... We shook hands with Lady [sic] Churchill ... She was smothered in wonderful pearls, few clothes. L'derry [Lady Londonderry] was wearing most of her heirlooms, looked rather quaint ... we stood around trying to look interested, then had some supper and were home at 12. Today I went to Queens, saw a degree conferred on Churchill. He seemed a funny little man, nothing much to look at...*
> [Elsa Loewenthal to daughter, Helen, at Bedford College, 3 February 1926]

Domestically she had a light touch.

> *You did not by any chance elope with father's new grey sweater. I bought it for Connemara and have turned out all his wardrobe today, but not found it ... Peggy's girl G[uide] Co[mpany] is to get the entertainment badge, she is to have a party here, But I am not to interfere or look after things, am only to lend them house.*
> [Elsa to Helen, 13 November 1926. Peggy, the youngest sister, was then 13]

Sadly, this attractive personality dissolved and became lost and Amélie, like Dixon, experienced maternal tragedy. While Amélie was a student, Elsa had a mental breakdown. She never recovered. To us, she was our second absent grandmother.

The letters provide documentary evidence, unknown until we read them – closely read them – that, without Elsa's younger brother, there would have been no Amélie in 'Hughie Graham's chemistry lectures', and thus no 'Amélie-' and no us; also, plausibly, no Cambridge job. We must digress, a longish digression, to explain that brother's role; the only male, apart from grandfather, JMcC, in our mother's side of the family.

First about him. He was 'Uncle Charlie', our great-uncle who lived out with vigour his sister's sociability (later he also demonstrated, like Elsa, mental fragility). As with our grandmothers, Uncle Charlie was known to us only in imagination and adult conversation. Unlike them, he sent presents. Thank-you letters, to 40 E 83rd St, New York, were an acceptable penalty.

Charlie had gone to that city early in the century, to establish a branch of Iklé Freres. He became rich and remained so, having presciently exited the stock market before the great crash of 1929. His sisters had failed to find him a wife:

> *Is it not hard lines that our only brother should still be a bachelor – when he has got three sisters always busy and choosing a wife for him. At the present moment he has got a Paris suggestion from me [Amélie's aunt, Olga], a Scotch girl recommended by my twin sister [our grandmother, Elsa] while my younger sister [Amélie's aunt, Amélie, for whom she was named] is much praising a Hamburg girl to him. And we are all equally unsuccessful – my husband [another cousin of JMcC] thinks it a fine change, to have me planning someone else's marriage. I think a man often remains a bachelor out of laziness or lack of courage to meet the responsibilities of married life.*
> [Olga Jacoby (née Iklé) to her physician, reprinted in *Words in Pain* [ed Catty J and Moore T 2019]]

She does not express the actual reason. Charlie was 'not the marrying kind'; the descriptor in polite society for a then 'pansy', now a gay man.

Charlie used his wealth for collecting and in patronage. His finest works were gifted to New York's Met but Richard has his Fujita, and Robert an 18th century crucifix labelled 'for Harry', the long-term boyfriend. Music was a special love. He befriended Ravel, a neighbour in Ciboure where Charlie had a villa, Horowitz played for Charlie and Charlie contributed to Prokofiev making his debut in Carnegie Hall, after which,

> *he went to Chicago in an effort to have his opera,* The Love of Three Oranges, *produced. To my surprise he was successful. And to my still-greater surprise he returned to New York a few weeks later*

and presented me with a reimbursement of the money I had contributed to his New York recital. To this day that occasion remains the only time that money I have loaned to an artist has been paid back.
[Quoted in Charles Iklé obituary brochure, 1963]

Of more direct relevance to our origins, Uncle Charlie's circle extended beyond music and the arts into science and medicine. Writing discursively to Amélie's father on New Year's Day, 1944, we find the relevant link. It is at the very end of Charlie's description of

a large reception at Sam Lewisohn's [distant relative]. They always have an interesting gathering. Among others were Dorothy Thompson [dismissive interviewer of Hitler before his rise to power, expelled from Germany once he rose], Wendell Wilkie [1940 Republican Party nominee for president], etc. The latter would be the only good substitute for Roosevelt, but his own party disagrees with him. So, I think he has no chance and I hope we will keep what we have. Tonight at 9 o'clock. Roosevelt is to speak on the Radio – It is always a thrilling experience to hear him talk – I am very interested to hear about Dixon's activities. I must tell Bobbie Loeb about it [sic] next time I see him. We had tea at his mother's last week and he looked old and tired.

Bobbie Loeb is the clue, distinguished physician at Columbia and author, with Cecil, of the dominant medical bible of his generation (and of Richard's and Robert's as London medical students in the '60s). Charlie was friends with Loeb's father, Jacques, a founding figure in biophysics. Bobbie's brother, physicist Leonard Loeb, was married, very briefly, to one Marion Hines who thus came, in 1924, to stay at the Loewenthal Belfast home of Charlie's sister, our grandmother. The Loewenthals lived in Lennoxvale, off Belfast's Malone Road, at number seven. The family was sufficiently prominent that, on envelopes, the number was usually omitted. (The less-grand could also receive letters addressed with little detail. A 1929 letter from Amélie, in Dublin, to *JD Boyd BSc, Greenisland, Co. Antrim* arrived the next morning).

Marion, of whom more later, was a specimen of that, then-rare, species: a female scientist. In 1924, school-leaver Amélie

4

was pondering her future. Writing to Dixon, while he was on a visit to the USA in 1930, she recalls that, when 18, she and her family took advice: 'if you happen to meet Marion Hines at J[ohns] H[opkins] give her my regards. She stayed with us once and was responsible for more than she guesses – she gave me the final shove into medicine.' Forty years after that shove, Marion 'remember[ed] doing my best arguing for you to be allowed to study pre-medical sciences, when it was history was it not that was being urged upon you?'. Thus encouraged, Amélie started at QUB and attended chemistry alongside Dixon, the first step to their future in a socially unlikely marriage. Marion and Leonard, by contrast, parted. Only his second and third marriages over the ensuing 50 years are mentioned in his American Physical Society obituary; history is selective. Uncle Charlie – gay, rich, intellectual, man-about-town – had, long before that death, unknowingly propelled the process of which this book is one minor culmination.

But, more, Marion also links Dixon to Cambridge anatomy. In 1924 she published a neuroscience paper in which she thanks the, then, Cambridge Professor of Anatomy, J T Wilson, for 'access to human material'. The visit to Lennoxvale must have been a side-excursion from that trip to Cambridge. Four years later, 2 February 1928, a paper on *The Brain of the Duck-billed Platypus* was 'read' by Marion at the Royal Society in London. She was 'introduced' by Wilson, world-authority on the mouth of such Australian oddities. When published, her paper was 176 pages long. (The material used for this blockbuster had been stored in the 'New South Wales Government Lunacy Department Laboratory'!). It seems likely that Dixon was initially brought to Wilson's attention by Marion. His first visit to Cambridge – 'It really is Cambridge, it really is' – was also to study that professor's 'material', perhaps of duck-bills. Lip development, as we will learn, was a major focus in Dixon's early research career. She may not only have contributed to our existence but also to that existence being in Cambridge.

We do not know what encouraged Dixon into medicine – perhaps his mother's consumption? Perhaps the example of a slightly older medical-student cousin, or of an influential adult? Stepmother Lily had, before marriage, been housemaid to a redoubtable Unitarian minister in Larne who seems to have become an important guide and grandfather-figure to the American teenager. Perhaps he played a role in the boy entering the Royal Belfast Academical Institution – Inst – Ulster's premier school, and thence QUB (Amélie's father, then unknown to Dixon, was chairman of Inst's board of governors). Veneration was accorded to that Unitarian minister but absolutely not to either Dixon's father or stepmother; rather the reverse. There is a whiff, in our father's attitude to the Greenisland family, of something more than understandable resentment at being pulled out of a transatlantic childhood or at the arrival of the Irish stepmother. Perhaps a financial betrayal contributed. Had his mother left money towards Dixon's education which had been used for other purposes? There were occasional comments to us which indicate something of this sort, and a few hints in the letters, but nothing definite. When Robert visited widowed Lily, years later, she claimed to have encouraged our father into medicine. To us, he never expressed any positive feelings about her or mentioned any such support.

The two school-leavers, about-to-become-medical-students, were clearly from very different backgrounds. The letters open our eyes to another difference. One between us and them. We, "Boyd Boys", were Cambridge children, Amélie and Dixon had been Belfast's. To move to Cambridge, they crossed not only a sea but also a cultural divide. In our childhood naivety, we never thought of them as other than entirely Cambridge. The Queen's University professor of economics, Hugh Meredith, another character as unusual as Uncle Charlie, and as socially connected, probably played a major role in their ability to navigate the intellectual transition from redbrick Ulster to King's Parade Cambridge. We will come to him shortly but first they must start at Queen's.

ii Medical Students

They were both still teenagers, 17 and 18, she slightly older than him. Despite Hughie Graham, any recorded intimacy was very gradual. An early hint is a somewhat formal 1927 thank-you letter for some student photographs:

> *Dear Dixon (hope you won't mind but Mr Boyd does sound rather overpowering!)*

Another, is the result of a wager on her BSc exam result settled at the beginning of the autumn term:

> *Received of MISS AMÉLIE LOEWENTHAL B.Sc. 1ˢᵗ Hons. The sum of 3 pence with required interest. JD Boyd with thanks 14.10.27*

It seems she passed; and lost the bet!

2. First name terms? *– 'Mr Boyd does sound overpowering';*
letter from Amélie to Dixon, 9 August 1927.

Once 'Dixon', there was progression in bridging their social divide and, at Christmas, he was invited to Lennoxvale where any dancing would 'not be too threatening'. She was now 22.

> *Dear Dixon, if you are in town on Sat 22ⁿᵈ – I will be so pleased if you can come to a little party at 8 o'clock. Yours, Amélie. PS very little dancing – just silly games, so please come!'*

Their evolving correspondence is more interesting about them than about their studies which were, we guess, diligent (at least Dixon's; his medals and prizes are listed in a later 'Medical Directory' entry). Intimacy was slow to develop; it took another year before, the morning after a January evening in 1929, 21-year-old Dixon was to write – the combination of emotion, intellect, reflection and, perhaps, overwriting is very Dixon (and note the role of the Unitarian minister, Mr Kennedy) – that his

> *cup is overflowing with so much joy that I simply have to write. After leaving you last night I was in such a reverie that I don't know how I got back to the hostel ... Because I couldn't settle down, I went to Greenisland. Conditions were as bad there, I couldn't sit still for two minutes and felt as if I wanted to shout "she loves me! She loves me!" to the whole world. So, I went out and was just in time to catch the last train to Larne. I arrived at the Manse at 11:30 pm and found Mr K[ennedy] still up. Here with him as confessor I managed to reach a sane though still very hyper-normal psychical state. I spent the night at the Manse but only slept in snatches – your wonderful presence was before me all the time. Oh Amélie I will do something to justify myself in your eyes. I remember Gregory in Science and Discovery writing something to the effect that a true scientist is devoid of human interest "like a Faust that has not a contract with Mephistopheles" ... I am like Faust who has met Margaret at the right time.*
> [Dixon at Townsend (student) hostel to Amélie at Lennoxvale, 20 January 1929]

Faust seems an odd role model. So does a night at 'the manse' for the later atheist chairman of the Cambridge Humanists. January 19 became an anniversary for them.

In 1929, Dixon was still bounded by the gendered fiscal assumptions of Raverat's Period, continuing that 'four years in some ways may seem a very long time but it may be a shorter time if I have any luck … yesterday when we kissed I felt that I had never been so near Paradise before. I really and truly mean this. Schuman's 'Traumerei' and Liszt's 'Liebestraum' have been chasing each other through my labyrinths since'; an appropriate inner-ear attribution for the, by-now, Bachelor of Science in Anatomy.

iii   Hugh Meredith – A Cambridge Apostle in Belfast

Ralph Meredith, a maths student with an aspiration to become a playwright was a schoolfriend from Inst. He was serious, intense, creative, and very left-wing, a political leaning then shared by Dixon. The friendship flourished and he also came to know Amélie who remained in awe of his seriousness. She wrote to Dixon from London some years later of going to 'a musical comedy with Ralph!! It was funny and we both laughed, but imagine going to that with Ralph – but he chose it … I admire his intellect and intense honesty a very great deal … Adam [younger brother, future professional Bridge-player] comes too … a nice kid'. Relations with Ralph were intense enough for Amélie to 'have the queerest dreams – such as Ralph making love to me'. Fortunately for Dixon, she was 'disappointed it wasn't you'.

It was probably through Ralph that Dixon and Amélie first got drawn into a student acting group lead by his father, Queen's economist, Professor Meredith. Hugh Owen Meredith was known to us and to others as 'HOM'; only occasionally 'Hugh'. He became and, unlike his sons, remained an important figure to both of them and later to us. He, as role model, seems likely to have implanted in the young Dixon and Amélie the later confidence that they could make Cambridge their own.

As Cambridge undergraduate at the turn of the century he had been an intellectual high-flyer. 'Without that long conversation with Meredith, walking again and again slowly around the quadrangle', writes George Moore of the final section of his *Principia Ethica* 'the central argument in that chapter could not have emerged'. Elected to the Apostles, a Cambridge indicator of brilliance (and source of Soviet spies), it was he who, in turn, proposed EM Forster's election in succession to Moore. Forster dedicated *A Room with a View* to HOM and remained a lifelong friend. According to the Dictionary of National Biography, 'it felt', for Forster, following a 1902 summer of intense friendship with HOM, 'as if all the greatness of the world had been opened up to him'. Forster would come to 'dinner' (rather than 'supper') when HOM – 'Hugh' to him – was staying with us.

After graduation, HOM had become Girdlers Lecturer in the newly established Cambridge Department of Economics, and a Fellow of King's College. But then he left (Maynard Keynes succeeded to his lectureship). HOM told Robert that the need to leave had followed 'an issue at King's with a married Fellow's wife'. His move was to Manchester University which had established a chair in economics in 1854, long before Cambridge, and where teaching of the subject was real world, part of a *B.Comm[erce]* course; likely attraction for socially-conscious HOM. It is more surprising that Manchester, with its strong Cambridge links – contemporaries in other subjects, Hill, Bragg, Rutherford and Dean, all moved from there to Chairs in Cambridge – was followed, for HOM, by Belfast where, somewhat improbably, this Cambridge Apostle had become Queen's University Foundation Professor of Economics. He remained there for 30 years and then, his first and second wives having died, moved even further from privilege to became warden of an offshoot of Queen's in Londonderry. Into his 70s, he there gave the young of Derry daily classes on commerce and more. He lived on-site, taking an active interest in their welfare. Socially-conscious Apostles could serve the under-privileged in ways other than spying for the Soviets.

Life in his Derry years was interspersed with vacation visits to Cambridge and to London where he met, and married, third wife, Peggy Spring-Rice, a London GP of distinguished lineage. They retired to her family home, Larkbeare, on Oxford's Cumnor Hill.

HOM and Richard

Early - me, in shorts, at his side as he strode across the great grass quadrangle beside King's Chapel, ignoring, to my anxiety, "Keep off the Grass". He reassured with, 'That doesn't apply to Fellows - once a Fellow, always a Fellow.' Later, in Oxford - he died quite suddenly and unexpectedly of a stroke there aged 84 in August 1964 - he wore a homespun shirt with a red tie, abandoning, as usual, the black velvet jacket (pipe in pocket) to the back of a wooden armchair. He was a Tolstoy, bearded and wielding scythe in the large wild garden of Larkbeare, who teasingly criticised an 18-year-old undergraduate's unwavering support for Labour. The economic policy of Alec Douglas Home (the 14th Earl of Home, renounced) might have elements better than that of Mr Wilson (distinguished wartime Board of Trade official and Oxford economics don). The erstwhile earl, with his Oxford 3rd class classics degree, admitted to using matchsticks to help understand financial questions but it was 'not a good thing' to have a PM who was the most able person sitting at the Cabinet table. 'Maynard Keynes was superbly able, doing, for example, what he did for Britain at Bretton Woods, but he would have been a terrible prime minister'. HOM wrote to me in his unmistakable green ink. A spidery hand but fun to decipher. I kept the letter. But, not in the archive, where is it now?

With the prejudices of the young, we were both surprised and intrigued to note, when Peggy and Hugh stayed, that only one of the two tactfully provided single beds had been slept in (Apostles were not all gay). The letters confirm it was a real love-match lasting until a note from Peggy, 'Hugh died suddenly today ... we were just enjoying our breakfast and reading our letters'.

HOM didn't just produce plays. According to *The Classical Review*, the 1937 translation into English verse of

'*Four dramas of Euripides* by [a] Professor of Economics should make professed students of the classics look to their laurels for they are the most interesting which have appeared for a long time'. His Queen's circle of student actors – James Bartley, Raymond Calvert, Bertie Rogers, and Ralph Meredith – drew Dixon and Amélie in a circle outside medicine. 'JOB [James Bartley – initials used in emulation of HOM?] … Raymond Calvert and I have formed a "Friday after-noon club" – meet in Union, drink TEA (truly tea) and talk'. Members of the circle became, like Hugh himself, extended family. Of them, the best loved was James, JOB. Failing to land a fresh Chair (he had had one in India) or a Cambridge Fellowship, James taught at Swansea University, alongside fellow lecturer Kingsley Amis, and wrote books. His last was dedicated to Amélie and Dixon (talking of dedications, our Christian – not, then, 'first' – names included Hugh for Robert and Adam for Richard; in honour, we were told, of Hugh and of Ralph's 'nice kid' younger brother).

James Bartley

The house lit up when James was staying. We were adults to him or so it felt. He was a year older than Daddy and, unlike him, a serious Ulster drinker. We, but not Mummy, were amused when the pair were 'tiddly' on coming through the door after an evening out. James's drinking was not restricted to evenings. When Dixon was dying, he came across the country to support Amélie and took her to a pub lunch. After food, and a few, he was flagged down by a police car. In court she was memorably quoted as having greeted the officer with, 'Thank God you've come.'

His friendship with Daddy was very Irish - multiple simultaneous implausible overlaps - the world of the mind and the world of literature. 'Did you thank Jo for *Ulysses*?' - the world of blarney, the world of empire, the world of scorn and the world of gossip. Despite the unsuccess of James's assumption that Daddy's pronounced academic progress would always be able, somehow, to draw him up too, their friendship thrived. James refreshed and nurtured our father's literary taste: total castigation of *Finnegan's Wake*; exhortation to see *Godot*; disdain for Virginia Woolf; admiration for *Hornblower* or for *Thy Tears Might Cease*; all extolled between the fourth and fifth Scotch in the pub, after a couple of glasses of sherry in the study, and with those slowly lifting clouds of cigarette-smoke.

Bertie (W. R.) Rodgers wrote poetry and became a somewhat improbable Presbyterian minister in rural Armagh. Later he was recruited to the BBC's Third Programme. Raymond Calvert penned *The Ballad of William Bloat* for a cast party. Its closing Ulster triumphalism - 'For the razor blade was German made but the sheet was Belfast linen' – might, after a good meal, be declaimed to us by HOM or Dixon, perhaps by both together. Two generations further on, it has been set to music and published with illustrations by Hector McDonnel. Its first stanza opens the initial meeting of the Dead Poets Society in the film of that name. Watching *Dead Poets* today, long after the actual deaths of HOM, Dixon, James, Bertie, Ralph and its author, where 'William Bloat' precedes a bardic cast of Tennyson, Byron, Shakespeare, Herrick and Whitman, gives HOM's influence an endearing extension.

Other students important in Dixon and Amélie's lives were very much not of the HOM circle. They included medical student Greenisland cousins, surgeons to be, Norman and Gerald, and two others we will also meet later. William Hamilton, in the year ahead, fellow medic and anatomist, later co-author and co-researcher, and John Doggart, engineer, Irish revolutionary turned Macclesfield manufacturer and husband of Amélie's dear friend-to-be, Sara (Sara was, like the Iklés, Jewish and John labelled his family and ours as 'Hibernian Hebrew').

13

iv   League of Nations Summer School

Dixon was something of a student-polymath. He read voraciously; books always a consuming passion. He became president of QUB's *Literific*, debated in Dublin, lectured in Larne on the *Celestial City* and was chosen 'to represent Queen's at a conference on *International Disarmament* to be held in the Lord Mayor's parlour' (four decades later he sent Stephen 'the Dufferin medal for Oratory, which your father gained because of his supposed debating skills'). Most formatively, he won a bursary to attend a League of Nations summer school in Geneva. It was 1929, the high water of aspirations for the League.

> *At Victoria Station ticket office I ... tripped over a little bag on the floor. It happened to be a portable typewriter and whilst making profuse apologies to the owner I discovered that he had in his hand a voucher like the one I had for getting a cheap ticket to Geneva ... Jenks is his name and ... he talks Cambridge English with a Lancashire accent. He comes from Liverpool.*
> *On the train we met another chap bound also for Geneva. He is Mossley who worked as a boy in the coal mines, emigrated to Canada, saved enough money there to come back to England and take a Tripos in Modern History. He, apparently, is one of the leaders of the Cambridge Socialist clique. And a very decent chap he seems. Jenks is taking a Tripos in Economics and International Law. They are both frightfully keen on the League and on social reform in all its various aspects ... they were surprised that I, as a medical student, should be interested in the League etc. They say that all the Cantab. medicals are quite disinterested [sic]. What a shame that we who follow a science which is also an art should get the reputation of being neither scientists nor cultured people? We (ie you and I) must never be like that Amélie.*
> Dixon in Paris, *en route* to Geneva, to Amélie, resident pupil Royal Victoria Hospital, 6 July 1929]

We find it impressive and, in retrospect, sad, how optimistically the League of Nations set itself to inspire the young with a high-pressure summer:

> *I have a discussion group in two minutes ... in nearly every spare moment that I have I feel too tired to do anything but sleep. Last*

week, in particular, I had to pay a very great deal of attention to lectures as they were on economics and, as you know, anatomists are not good economists and it "fauted" me to devote a good deal of spare time to self-examination in matters economical and to getting Mossley and Sloane to explain difficulties. Sloane is a terrific "brain". He has a first in economics (Clare College Cantab) and [is] very willing to help. [The Clare magazine 92 years later features a student attending the UN in New York on a similar summer bursary – nothing new] ... The lectures are very good. J.M. Keynes was with us and a very great man he is. He was the brains behind the Liberal Party at the last election and is probably the greatest economist in Great Britain. During the war he was in charge of the dept. of the Bank of England controlling foreign loans and at the Peace Conference was one of the British representatives. He resigned because he felt the reparations problem was being tackled in a very bad way and has always maintained that the saddling of Germany with the immense war debt is not at all for world harmony. Through my intimacy with Jenks, Mosely and Sloane I was invited to Keynes house for supper on Wednesday evening. It was very interesting. K. himself is married to a little woman about 1/2 his size (he is quite tall) who was a very famous ballet dancer – a pupil of Pavlova herself – they were all very jolly and when she said something about not agreeing with him and wishing to ventilate the question he said, 'Yes I am sure you could give quite a good "legture".'

The other lecturers were Prof Bonn of the Berlin Hochschule and Prof Rueff of Paris University. Bonn lectured in English which he speaks very fluently, Rueff in French ... much easier to follow the French now ... one night we had Prof Brunschweig [actually Brunschvicg] of the Sorbonne who after Bergson is probably the best modern French philosopher. He lectured on the idea of universalism applied to the concept of internationalism ... On Thursday I had the privilege of hearing [sic] a sitting of the international committee of Intellectual [?] which was sitting the whole of last week. Z[immern] was presenting a report on some work he has been doing with regard to the teaching of the League in the various nations and he managed to get some of us tickets of admission to a private sitting. It was very interesting to see some of the most brilliant men in the world met together in solemn enclave to discuss the working together of intellectual conceptions and endeavours of the various nations into ideas of international value and significance. Gilbert Murray was chairman of the committee ... He is a great man.

[Dixon in Geneva to Amélie, Resident pupil, Royal Victoria Hospital, Belfast, 8 July 1929]

They also unwind hard:

> On Friday there was a dance at the Hotel des Bergues and at
> 7.00 A.M. a whole crowd of us went, first to a whole night "brasserie"
> and got some wonderful food and good beer and then walked up to the
> top of Salève in our war paint to see the sun-rise. Lord but my feet were
> sore but it was worth it. Another night, after the opera, we went to a
> café near our hotel in Montparnasse and we sat (the two Cambridge
> chaps and myself) until 2.30A.M.(!). A wonderful experience it was
> with the exotic sounding drinks, the waiters in dress shirts and trousers
> and little white dissecting [says the medical student] jackets, the
> Bohemian looking men with long hair, huge bow ties of soft silk
> hanging down to their waists and artistic looking faces ... I had a letter
> from Ralph Meredith. He is due here sometime next week. He says he
> is having a wonderful time enjoying his walk across France thoroughly.
> I am looking forward to seeing him as I like him very much.
> [Dixon in Geneva to Amélie in Belfast, 30 July 1929]

There is a hint, when Dixon returns to Geneva two summers later,
that the League's 1929 idealism is beginning to seep away:

> Last night I attended the Socialist Club at the School of which I am
> a foundation member. We met (as of yore) in the Café Landolt where
> Lenin and his confreres plotted the Russian revolution. The present
> set have different ideals from those of us of two years ago and there
> were only four foundation members present – Sloan (yes Pat Sloan
> is here before going to Russia [where he spent most of his life]), an
> American called Stillman, Liesel Klaus [?] and myself. I took [two
> illegible names] along and they were really very interested.
> [Dixon in Geneva to Dr Amélie Loewenthal, Royal Victoria Hospital,
> Belfast 4 August 1931]

Amélie's reply is, though not yet a Boyd, more wifely:

> I don't see how you could help attending your Socialist club. I expect
> you strutted about with your chest stuck out supporting that ghastly
> grizzly red tie – not to mention the superb green beret – can't I just
> see you staggering up to Chamonix in it – oh well I'm sure it will look
> lovely against the snow. Anyway, I wish I were there with you beret
> or not. What fun meeting so many of the original gang again.
> [Amélie at Royal Victoria Hospital, Belfast, to Dixon, Pension des
> Bostious, Geneva, 10 August 1931]

But the 1931 visit was not just social. The young boyfriend, now Dr Boyd and junior anatomist at Queen's, had international aspirations and took advice. Openings at the League might be in

> *the Section of Industrial Hygiene at the B[ureau] I[nternationale de] Travail [or] the Medical Secretariat of the S[alle] des N[ations]. Zimmern has spoken to both about me and I have an interview with the chief of the latter on Monday. Of course, I will not get anything definite for years (if ever) but I don't think I could ever pass over a League job if the chance did arise in the remote future. Jenks assures me that he knows of no possible vacancy in the Medicine section for at least five years.*
> [Dixon in Geneva to Amélie at Royal Victoria Hospital, Belfast, 7 August 1931, pencil]

Zimmern was something of a visionary. With his wife, he ran (and had founded) the Geneva School of International Studies. Earlier he had drafted, while working in the Foreign Office during the war, the blueprint for what would become the League of Nations – 'a regular conference system ... open to universal membership'. Shortly after the interview Zimmern had arranged, Dixon received 'a note from the Swiss chap who is in the medical secretariat. He wants to know if I would like to go to China if the L[eague] of N[ations] send a medical relief expedition ... Maybe I <u>ought</u> to say <u>yes</u> but it would wreck Anatomy for ever'. Despite the earlier protestation, Anatomy won. He did 'pass over' the opportunity.

That the League's failure to prevent the invasion of Manchuria by Japan was damaging was already clear to Amélie. 'It is dreadful that the League has lost so much cudos [*sic*] over having so ignored the East'. Her father is slightly more optimistic. 'As for Japan I should have liked the L of N to take a firm stand, but perhaps it will first limit and then end the fighting'. The rest of the decade would sadly illustrate that such optimism was not to be justified. Dixon correctly pointed out that

> *the Manchuria dispute is fraught with frightful consequences to the L de N ... It is a problem when a young country with an intense <u>and</u> growing nationalism like Republican China comes into close*

economic relationship with an overcrowded people who have an
Imperialism as blatant as the Chinese nationalism is rampant.
[Dixon to Amélie, date uncertain]

v   Graduation and the Road to Marriage

In June 1930, Dixon and Amélie had taken 'finals', MB, BS,
BAO. 'Best of luck, Amélie, all through it. But remember "these
things too shall pass away" *et puis*!!' They become 'doctors'.

Friends and relatives had rallied to support Amélie before
the exam. Joan's advice throws an interesting light on women in
medicine.

> *Put on a nice dress for the oral and perhaps an engagement ring, like*
> *... the girl who was taking her M.B. exam and seemed rather at a loss*
> *– then the examiner noticed she was wearing an engagement ring*
> *and found out she was going to be married in a fortnight – so he*
> *let her through because he thought she couldn't do much harm in*
> *that time.*
> [Joan Loewenthal in Paris to Amélie]

More than 40 letters, telegrams and congratulatory phone calls
(jotted down by her father) are from friends and fellow students
near at hand and from family as far afield as Frankfurt, Paris, and
St Gallen. Thus lauded, she headed for a summer of relaxation on
'the continent'. For Dixon there must surely have been at least a
handful of congratulations but none remain. He would have been
pleased that the professor of medicine

> *regard[ed] Dr JD Boyd as the most brilliant student of his time at*
> *Queen's University ... it is only necessary to add that his year at the*
> *University was a very good one.*
> [Handwritten reference from Thomas Houston, 2 July 1930]

Dixon headed off to the States where he camped with cousins – a
most un-Dixon-like activity – and visited Johns Hopkins (and
Marion Hines). Was he job hunting? Travel documentation lists
him as 'US citizen planning to stay'. That seems unlikely to be
the truth. Amélie writes to him from Uncle Charlie's Ciboure

Villa where she is enjoying the lifestyle. (At which villa Robert was, in August 1937, allegedly conceived – a holiday from which, perhaps fortunately, no letters have come to light).

> *Had lunch yesterday with Maurice Ravel (in case you don't connect him up he's French modern composer). It was rather fun because the wee man never stopped talking for five hours – but scintillating with wit and such acutely intelligent conversation he was like a goldfish going after a crumb! – darting after any stray remark ... [I will] probably stay here ... till end of August, then motor with Ch[arlie] to Switzerland and see the Fritz family [Iklé cousins] and Mother if it is thought possible [Elsa is now a patient at Pranguins, a psychiatric clinic near Lausanne], stay there a few days and come home mid-September; that gives me a fortnight of being a "good daughter" before October 1.*
> [Amélie to Dixon in USA at cousin's New Jersey home, 28 August 1930]

On 1 October 1930 they both start work as junior doctors at the Royal Victoria Hospital, the RVH. She 'on wards 9 and 10, he on 5 and 6'. Like most new medical graduates, she is somewhat nervous – 'Have forgotten so much already; you will have to help me'. They will have been busy.

3. House physician – Amélie in 'her' ward with nurses, Royal Victoria Infirmary Belfast, 1930.

Seeing each other daily at the RVH, there is no correspondence. The following year she took an obstetric post in London and there are almost daily missives. He, the League of Nations declined, had become 'demonstrator' in the Queen's Department of Anatomy, teaching medical students and working on an MSc project: *Development of the Lip*. His redoubtable Professor was 'Tommy' Walmsley. At least four of Walmsley's students became anatomy professors, so did Walmsley's younger brother. The letters' dominant theme is love, interspersed with comments on work and reading, art and music, and on family. Sometimes Dixon writes from a Liverpool hospital where he does holiday 'locums'.

She [15 July 1932] has 'lived more sincerely thro' loving you than for all the rest of my life together ... all this is really quite simple. I love you back! My letters get ungrammaticaler and ungrammaticaler – but you can't write love on Fowler [*Modern English Usage*] – oh damn'. He writes [31 July 1932] of being 'maudlin with sleepiness but still compos mentis enough to know that I want you to be with me <u>always</u> and <u>always</u>. It is my only surety in life. Without the hope and the wish of being married to you I would be a sorry fellow'.

In his ecstatic letter after their 1929 January evening, Dixon had contemplated a four-year engagement but, for the young doctors, desire appears to have trumped or come close to trumping matters after he had, on a visit to Greater Manchester, 'wired' to seize an opportunity:

> *Post Office Telegraph (30 Mar 1932. 2.50 Salford to South London Hospital, Clapham, London. Time of receipt 3.20 COMING FRIDAY RUGBY = DIXON*

From the boat back to Belfast he complained that

> *it took <u>six</u> hours to remove the evidence [after a moment which] seem[ed] like a 17th. century idyll ... or a second Daphnis and Chloe ... my being with you <u>in a hay stack</u> ... and actually falling asleep in your arms.*
> [Dixon (on 'LMS Railway Company - Heysham and Belfast Steamers' paper) to Amélie, 4 April 1932]

Before the next move in the relationship, a summer intervened. Summer activities point up their different circumstances. He sailed across to Scotland with fellow students, now young doctors, in a student boat, *The Surprise,* bought by the group for, allegedly, £20. She holidayed more spaciously in Italy with sister, Helen.

> *We are leaving [Venice] next Wed 14ᵗʰ then Padua for 2 days and then [Switzerland]. Will you send me your address there c/o Uncle Fritz ... Herr Fritz Iklé, St Gall.*
> [Amélie, Hotel Parco, Venice to Dixon, Walton Hospital Liverpool, 6 September 1932]

No street address was needed for the home of a second-generation Iklé Frere, one of St Gallen's largest employers. Such different circumstances were, we can surmise from a comment of Dixon's, occasionally a difficulty.

> *Gossip is the only thing one can really laugh at – even when it is malignant. You should have heard me laugh when a person ... congratulated me on "managing to become engaged to a very wealthy girl"!*
> [Dixon at Greenisland (Salford Hospital paper) to Amélie, South London Hospital for Women, 11 April 1932]

Amélie's sister, Joan, made a point of ignoring the difficulty:

> *At Euston who should I meet but Miss Joan Loewenthal. I was trying to make my mind up about a novel for the train and someone suddenly shook me by the hand. It was Joan, looking very well and acting very pleasantly towards me. She is very sympathetic towards US [his capitals], 'melie [Amélie], and I do love her for that.*

Class managed; a night together could be too. Dixon's student nickname was 'The Bishop' and, signed in Dixon's hand as J.D. Bishop esq., a Stratford Hotel receipt confirms pre-marital cohabitation. We wonder whether Amélie wore a ring to get past any suspicion from Miss Cran and Miss Watney, the hotel's owners whose names are printed at the top; 'esq' will perhaps have helped. The Shakespearean choice of location is characteristic of them both and the retained receipt suggests happy memory. The next day is his birthday and she wishes him 'many happy returns of tomorrow and yesterday'.

But love is not smooth. A rebound later that autumn swerves the relationship close to destruction. A single letter survives in which Amélie writes decisively, and indecisively.

> *I have been trying for weeks to write you a letter – that I would not see you again, at least for some time. That the strain, which we impose on each other is too great. Each time during the last weeks when I have left you, you have been unhappier and more disturbed, it seemed to me, for having been with me. [But] ... you have confirmed my sense of values many times, and I have felt the most certain happiness and contentment at moments in these 3½ years.*
>
> *Well then – I would rather not see you again, at least until next year, that's all.*
>
> *And now to be practical [typical Amélie]; I am seeing Grace [mutual acting friend] on Thurs. as arranged, I will make an efficient excuse for you – so that she won't attribute your absence to any rudeness – you weren't well. I am going from the Quad at 6:20 ... Darling, please don't write ... And Dixie [again, typical Amélie] – I do think you should have that darned tooth seen to - and stop reading at 12:30.*
>
> [Amélie to Dixon at Greenisland, 16 November 1932]

The injunction 'until next year' is not obeyed. Positive letters are flowing both ways the following week but, after Christmas, things are again uncertain when, in a water-damaged patchily-illegible letter, Dixon offers a pause and a let out.

> *Do be good to yourself in the agreed few weeks and be your own soul's[?] arbiter always. Already I yearn to hear your voice again (yearn is the only word). Please 'meliest [Amélie] consider well in the next few days our whole affair dispassionately and away from me and make [?] the absolutely ultimate judgement. I love you wildly and without reserve of any self-dignity but ... it is your choice that must be the true judge.*
>
> [Dixon to Amélie at Lennoxvale, 5 January 1933]

In next day's reply Amélie's choice is clear – marriage. And how it should be viewed.

> *All say it is a mistake and a handicap for a man to marry when he is so young – under thirty, the argument being, I think, that he has not had enough experience of other girls and that a wife is an unnecessary care and absorbs time and energies which should be used for "getting on". I think this is only a snag if one is not aware of worldly values so that one can disregard them. Neither am I a practical wife for you, never having had the advantage of having to cook my own meals. (I am much less pessimistic about this – the Technical [College] is going to be a tower of strength to me!) And age [Dixon was 17 months younger] shouldn't matter with us, for you have always been older in experience of living than I. And it will be far nicer for the kids to have a young father! But it's all negligible compared with the longing I have to be with you and the peace I get from it – just your company. I can only have that for always if we marry. Darling, I promise you "dispassionately" that I have believed us to be right and that I am sure our sort of love is the salt of the earth – your letter was so lovely and honest. You mustn't be terrified. I am "jumping in" with my eyes well open and I too love you so. I'm afraid you will be having an unsettled week.*
> [Amélie in Edinburgh (visiting her institutionalised mother, Elsa) to Dixon, Greenisland, 6 January 1933]

The wider world was unsettled too. Unmentioned by them, Adolph Hitler – Make Germany Great Again – had, by the month's end, become chancellor.

The suitor would need to speak with potential father-in-law, JMcC, about marriage, so Amélie broached the issue and was pleased to convey

> *how completely sympathetic and altogether understanding my father has been to us. He told me that it had not been through lack of interest that he'd never asked me about you, but from the dislike he had of interfering in anyone else's business, that you wanted me to be happy and that therefore it was a question largely for me to decide (and you know all about that). Then that I should have an income of my own,*

but that he thought it right that a wife should help her husband if she could and that we should always 'live within our income' [very JMcC] ... he seemed to me quite pleased. You won't find it hard to talk to him now I <u>know</u> ... it would have been hateful to have had to react to either his or Tommy's disapproval. Both Tommy [Walmsley] and Daddy behind us – and I sure of you – and you of me; we do start with a fair wind.
[Amélie in Edinburgh to Dixon in Greenisland, 25 March 1933]

'It is very difficult', replies Dixon, 'to say how great a difference is made to me by your father being sympathetic. I have always felt most acutely the semi-shielding that we had to do'. Now, Amélie points out, Stratford strategies can be left behind. 'Poor Mr Bishop late lamented. I feel deeply grieved at his demise; overwork I suppose, he was inclined that way, though sympathetic and charming and liveable with – so they say in Stratford, and with such an excellent taste in pyjamas. Someday we must have a pilgrimage to Stratford and visit his grave'.

Marriage finally took place in August 1933, but not without a further, to us astonishing, swerve. Amélie once wondered, disconcertingly to semi-adult Robert, whether her failure to bear a daughter had been 'God's punishment' for her having had an abortion. They were just leaving the theatre after seeing Ralph Richardson in *Flowering Cherry*; he was 19. It was never mentioned again but letters suggest that she almost certainly had that abortion three months before the wedding. Perhaps the memory was brought to mind in Shaftsbury Avenue because Robert was, on parental advice, about to consult a turn-to medical relative, psychiatrist and distant cousin, Eddie Bennet whose Harley Street location was not far from the theatre. Twenty-four years earlier and only three months before the wedding he provided medical cover for her having a 'D and C', a procedure of uterine clearance, legal for menstrual irregularity but then criminal if used to abort. Eddie was also involved, professionally, in the care of Amélie's mother, Elsa, and, sexually, with her sister, Helen. Complicated!

The D and C appears to have been Amélie's unilateral choice. She had been at a conference in Dublin when from the Dublin-Holyhead boat she writes ['posted at sea' 1[?] May 1933] to inform 'my dearest Dixon ... there is no possibility of my getting back for Thursday. It is all very confusing, but I have had to go on to London. Sounds like an Edgar Wallace! However, I will explain the story to you when I get home. I promise'. Dixon followed her to London and there is an empty envelope inscribed: 'This letter destroyed on the 8th May 1933 [her 27th birthday]. Reason well known to us both. JDB'. In a further letter to Dixon, now back in Belfast, she advises on secrecy and on contingent cover stories and also answers (her capitals and punctuation) the obvious question in just two, curiously located, capitalised words. She is

> *absolutely marvellously recovered – slept beautifully and have a pulse you would be proud of ... Do hope you got back comfortably and didn't walk into T.W. [Tommy Walmsley] on the platform – I had visions of that. I can't tell you what a comfort it is to me that <u>you can tell a</u> good lie and probably that is the "explanation" you want to know sometime, of why it was ITS YOU! ... Let me know what Peggy said ... do you think it would be a good thing to tel[ephone] again to tell her that you have heard from me ... and that I seem to have had a busy time – or something to that effect ... D[addy] unfortunately met [fellow junior medic] Spud ... [who] asked for me and D. told them, I am sure, that I was in Dublin at P[ublic] H[ealth] congress and reading for the exam [postgraduate Diploma in Public Health] – which Spud won't easily swallow being one of those awkward intuitive people. Perhaps that will mean some more inventions for you to work up – anyhow I warn you! He will put 2 and 2 i.e., your departure and mine together – so? "I wasn't well, came over here to see my Dr. cousin – didn't want the family to know etc. etc. etc".*

[Amélie, recovering from abortion, at Castle Hotel, Deganwy, N Wales to Dr Dixon Boyd, Queens University, Belfast 10 May 1933]

Invention wasn't necessary. 'I breathed much more freely when I'd read your "all clear" ... Glad it was unnecessary'.

Only a sole letter in the remaining three months until the wedding mentions these, to us, strange two weeks in early May.

This morning I was in Harley St. It did make me laugh to think of my last meanderings [there]. I long for me to be having a child that's growing from you and me and for you to be looking after us. Last time even that tiny baby hurt me so – and that even then was all part of loving you. So, think next time when it's a living soft baby of ours that will grow.
[Amélie at 34 Albion St. W.2 (Sister Helen's flat) to Dixon, Walton Hospital, Liverpool, 27 June 1933]

'Part of loving you'? To protect Dixon from being labelled (in the phraseology of the time) a 'shotgun' husband is our puzzled interpretation of those four words a near-century later. Would the unborn have been the missing daughter? There is no evidence that her entirely 'woman's choice' in how to proceed was ever resented by Dixon, but 'laugh' does still grate.

Dixon doing a summer locum at Walton Hospital in Liverpool is in pre-wedding-conventional mode. He braves JMcC's sisters: 'I am glad the "Aunts" were not troubled by my letter – I had more temerity than I should have had – for several days I was quite convinced they would be annoyed'. In fact, they responded with positive practicality, giving 'nice and useful furniture for our flat out of their surplus – viz: 2 wardrobes, bed and mattress; 2 rugs; armchair; 6 chairs; chest; curtains; blankets; cutlery (1 doz. of each) and a fish slice which belonged to my great Grandmother'. Reports of presents and planning come on apace. 'The Bennets [he of the 'abortion', 'termination' a later euphemism] are giving us blankets – nice coloured soft ones'. Elsa's sister, 'Auntie Amélie', apologises from Hitler's Hamburg for delay in sending a tea set. Lady Livingstone, vice-chancellor's wife and neighbour in Lennoxvale provided another, albeit a morning one, 'which from Peggy's description "blue with parrots" sounds rather hideous. But I suppose one doesn't feel v. aesthetic in the early morning anyway' (We boys used to love breakfast-in-bed, an Amélie special treat served from that set, one parrot hidden by porridge with cream and brown sugar, another perched on the teacup; crisp bacon and fried egg; toast and marmalade).

There were many thank-you letters to send.

Meant to tell you yesterday not to acquire another wardrobe fr[om] your family as we now have 3 – in fact wardrobes have outgrown all other furnishings ... Seems a pity if we have 2 lots [of] ... silver tea pot, and cream and sugar ... [when] we still want: fruit knives and forks, or sauce boat; or muffin dish: we don't have to have any of these things, but they are useful [Really? – the Greenisland boy was certainly moving up a class] – and it's all we haven't got.
[Amélie in Edinburgh to Dixon, Walton Hospital, Liverpool, 11 and 20 June 1933]

Dixon was immersed clinically but does find time to comment. 'The invitation does look well; I am terribly proud of it. If you have any left-over, would you send one to each of the following [medical staff at the Walton] ... They have been terribly decent to me'. Elsa's possible attendance was more fraught.

Mother [Elsa] is a shade better, but I think coming home is out of the question ... D[addy] was so determined to bring M. back ... He wanted her to be there for my wedding too, and poor M. is too ill to want to know what people round her are doing.
[Amélie in Edinburgh to Dixon, Walton Hospital, Liverpool, 12.7.33]

Dixon empathises but is realist – not only is it Amélie's call but they need to look forward.

I had a feeling that I should like your mother to be present at our wedding, a sort of wish that it might do her good – but I am sure it is much wiser otherwise ... Honestly though, 'meliest [Amélie] the past is the past and with all my heart I wish you not to worry[?] but to look forward – we are young and the world can give us a very great deal; other people are other people even if they are one's mother and father. You know darlingest that what I write is neither callous nor an attempt to explain things away. I want you to be as happy as anyone ever has been on this too, too, solid and that can only be so by doing your duty as you see it not as anyone else does.
[Dixon at Walton Hospital, Liverpool, to Amélie, 21 July 1933, water-damaged]

At times the preparations seemed a little absurd. 'Next time we get married, we will go to a strange town, drop in on the registrar of marriages and that will be the end of it all', says Dixon. Amélie's concept is slightly more elaborate.

> *Just uncle Fritz [St Gallen Iklé cousin of Elsa's] and HOM and Miss Praeger [sculptor family friend] and May Earles [unidentified] there beside ourselves. I wonder who you'd ask. Yes – Madame V. [unidentified] and Mr Kennedy [Unitarian minister] I'd have wished for too, and Mother – and we'd have had a picnic on Colin [Co Antrim] with Chianti and there would be a rainbow – and then suddenly it would all disappear and you and I would be walking down the Champs Elysées at about 10 o'clock at night – and Paris lit and fairy like!*
> [Amélie at Rock Cottage (Loewenthal Co Down hideaway) to Dixon at Walton Hospital, 9 August 1933]

Elsa was not there, when they married at Belfast Registry Office on 21 August, but many others were. Witnesses were Norman Townsley, Joan, JMcC. Fathers are given as James Dixon Boyd, Grocer ['grocer' seems a bit of a stretch] and J. McC. Loewenthal, Linen Merchant. According to Helen, 'it was a stupendous success. Am off to Miss Milner's to give her list of cakee's'. Norman was best man.

Presumably other members of Dixon's family were also at the wedding but, of them, there is no mention. Norman was Dixon's first cousin, older brother of Gerald, both sons of Dixon's aunt, Annie. The boys lived round the corner in Greenisland, Dixon between them in age. In youth, he sometimes slept there sharing a bed with Norman (a fact only mentioned when Robert opened the topic of adolescent sex – 'nothing like that with Norman' was the slightly huffy comment).

vi   Two Surgical Cousins

The Townsley cousins were, like Dixon but more directly, of the rural Antrim working class and similarly, through QUB and medicine, left that heritage behind. Despite probably being born illegitimate, they became well-regarded, more than well-regarded, provincial surgeons in Norwich and in Kent. Evidence on their origin is confusing. In the 1901 census, Anne and our grandfather, Dixon Sr, are listed as living with five more siblings in their parental home in Straid. By 1911, Dixon is gone, living in America with Grace and young Dixon. Annie is gone too but two Townsley children, Thomas and Bernard, are with the grandparents. There was no 1921 census – the civil war – but by then those same two, now bizarrely renamed Norman and Gerald, are with Annie in Greenisland where, to Robert's insecure memory, no Townsley father or stepfather was in residence.

Both were later to marry not only interesting wives but also ones, like Dixon, who were born middle-class or more. Norman's Alice, from a distinguished medical family, was, like Amélie, an early female medical graduate, in her case from Glasgow (one of only two in her year) where grandfather, Sir William Gairdner, had been on the staff. Unlike Amélie, she became a consultant, in obstetrics. Gerald, in late career, married 'his' theatre-sister whose father was vice-president of Luxembourg's parliament.

Norman, as best man, worried about his speech and about his morning coat. This somewhat gauche, deeply Ulster, semi-brother chosen for that role ahead of William Hamilton, or James Bartley, or Ralph Meredith was always, by our Cambridge family, covertly condescended to. Amélie remembers 'what romantic fun we did have but I can't endow Norman in those days with any other attribute than that of "Pardon"'. Perhaps the condescension came from her – 'What a funny, kind, conceited, blinkered creature he is'. Norman and Alice were always lovely to us. Medical student Robert, privileged to stay and warmly, very warmly, welcomed to Norwich to see real surgery with Norman, was told that catgut knots should be tied 'the way the boss likes them tied'. To describe Norman as 'boss' seemed to him ridiculous. He had, it seems, adopted Amélie's perspective.

Richard at the Townsleys

Boxing Day, in the early 1950s, involved driving in the Hillman Minx from 48 Grange Road to 97 Newmarket Road; from Cambridge to Norwich; from what now had become one cousin's upper middle-class world to another's with a different tone. Daddy had had financial advancement through marriage, Norman's followed from a flourishing surgical practice in a well-heeled provincial city. 'Look under the bed, Dixon. Those cardboard boxes have the Georgian silver. And they are completely full ... I'm now collecting watercolours by the Norwich school of artists. They're quite nice and very good investments.' Both true. Roedean for Janet and Westminster, for young Gerald, followed.

Boxing Day lunch - quite delicious - was fricassee of yesterday's turkey. And Norman, over dessert, always showed us, hidden in his waistcoat-pocket, the scalpel ready to be wielded in case of need. Need, in those pre-defibrillator days, being the ability to open the chest and massage a fibrillating heart. 'I know that I may be called on to use it one Saturday afternoon on some ticketholder at Carrow Lane. The Canaries are having another unsuccessful season this year.'

Gawky, smaller and more bent than the very upright Alice, Norman had an Ulster twinkle in his eye. He was kindly. He, and Gerald too, were, we believed, somewhat in awe of the academic path of their 'super-brilliant' Cambridge cousin.

Norman advised me (me a mere clinical medical student, he the senior surgeon to the Norfolk and Norwich) in a confidingly affectionate whisper across the operating table: 'Richard, if you ever have the chance to sew up for the boss, take it. That's how I, unlike all the other boys at Salford Royal, gained a starring reference after my three Manchester years from the great Geoffrey Jefferson. And that put me on the road to becoming who I am now, the surgeon here.' All this with the twinkle. The fantasy of meritocracy must be trumped by who you know - an inevitable fact of life.

Why the Boyd condescension? – academia over practicality; a university stipend rather than fees from trade in the rough and tumble of nursing-home surgery; orthodox childhood over bastardy; envy of part-time Amélie for a real consultant – we remain puzzled but are left feeling slightly soiled by our collusion. Others remember their warmth too. An 80-year-old friend of Richard's recollects Norman's kindness to her as a junior theatre nurse; a 'family-planner' casually met in recent years describes the admiration she had for Alice.

Both Townsleys had risen to become surgical leaders. As juniors they had visited outstanding European centres. The Gerald Townsley Trust founded by colleagues in Gerald's memory focusses, appropriately, on travelling bursaries. Norman – Larne Grammar School, not Inst – became a member of the Travelling Surgical Club. That was prestigious and he was secretary on its first post-war visit to Norway where he had landed in 1945 as Lt Colonel Townsley, medical officer to an airborne task force. Dixon didn't have that.

Four years' challenges of class, age, sex and social mores having been trumped by their marriage ceremony, the couple departed for, indeed, the *Champs Elysées*. Honeymoon-Paris is followed by Spain and Majorca and return to married life in 4 Queen's Elms, Belfast, the first of ten homes they will share during 33 years of marriage. Amélie's sisters have organised things while they were away. Helen reports progress and gives advice.

Joan says the flat is fine and almost finished now. I have not seen it for two days. Alec [Lennoxvale chauffeur extending his role] is really ecstatic about all his work. And you'll have to admire every square inch I warn you. Not to mention things like the bathroom cupboards done by the Galbraith boy.

As 'cook general' I am so glad you are not huffy about Mrs Shaw. You really ought to be very glad as she is <u>dreadfully</u> superior ... You'll have to be a bit graceful with her. Of course, the more work you can give her the better as otherwise she will have to take cooking jobs or something in the afternoons so that there won't always be someone to

answer the door ... Give her 10/ a week and she'd stay in all afternoons
for that. She'll always do washing, baking etc. and dinners ...
[Helen Lowenthal to Mrs Dixon Boyd ...Thos. Cook and Sons Ltd
... Madrid, 11 September 1933]

Junior-doctor life, 1930s Belfast-style indeed. Settled in Queen's
Elms, there are, again, no regular letters that year. There is only
one, from Amélie, *en route* to an Iklé winter holiday in
Switzerland. Her love of travel and Dixon's more sedentary
preference became a recurrent theme.

vii Cambridge Job Offer

At the end of 1934 came the invitation to a Cambridge post. Dixon's
work was going well when he received the New Year's Eve letter
from Harris, new Cambridge anatomy professor in succession to
Wilson (mid-century, a professor was 'the professor', appointed as
head of department, and of 'the discipline', for as long as he (*sic*)
was in post).

Walmsley, in a handwritten note congratulating Dixon on
a gold medal for his lip thesis, is supportive of the offer:

> *My dear Boyd,*
> *I have thought over the matter of your letter and compared the post*
> *you are offered with the possible vacancies as well as I am able; and*
> *I think you should accept it. It is a sure thing to come back to [Dixon*
> *had received a Rockefeller Scholarship to fund a year at the Carnegie*
> *Institute of Embryology in Baltimore, USA]; it is a fair salary (at*
> *£450); and it is a good chance, for in spite of the demonstrator status*
> *the openings should be good and, in a year or so after you go a*
> *re-calculation can be made ... I take it Mr Lowenthal [sic] is*
> *agreeable.*
> [Prof T Walmsley in Kirkcudbrightshire to Dixon at Rock Cottage, 19
> July 1934]

Interesting the emphasis both on money – a recurrent Dixon
worry over the years despite Amélie's origin, perhaps because
of it – and on her father's view! Harris's letter, quoted earlier,
continues, making the opportunity sound very attractive.

... your teaching hours will be UNDER 16 hours per week and you will have such vacation as you have never dreamt of for the term here is only eight weeks.

Walmsley was here as Examiner for the 2nd MB Exam. He had a good view of the place – and of how much was to be done. The authorities here are treating me well and this indicates that we can make a real department – with young blood and fresh outlook.

Give my kind regards to Streeter, the Lewises and the various colleagues [at the Carnegie Institute].

I remain yours sincerely

HA Harris

I would like to receive [your details] as soon as possible so that the Appointments Committee may ratify your appointment as University Demonstrator.

Ratified it was – appointment committees, then as now, often ratify rather than appoint. Before Cambridge, there is the year in Baltimore.

1934 was an academic *annus mirabilis* for Dixon. It is 'exciting to think that you have had now two big articles and three shorter in the Anatomical [Journal of Anatomy]'. His thesis on development of the lip – so important for the ability to suckle; and for his career – was well regarded. Amélie, in London, hopes it will do well:

if it is a g[old] m[edal] in June won't it be thrilling [it was] – that and the Rockefeller in six months – won't I just be proud to be Mrs.J.D.B.! What fun that Le Gros Clark [Oxford's newly appointed professor of anatomy] has appreciated your 'lip'. It really is marvellous Dixie.
[Amélie at Ballycastle to Dixon at 3, Queen's Elms, Belfast, 12 April 1934]

viii A Year in America

Amélie, potential academic, is that year still thinking of herself as scientist as well as wife.

Carrel [1912 Nobel prize for experimental surgery; an early user of tissue culture (T.C.)] T.C. is awfully interesting. I finished the 6 articles in the Journal of Experimental Medicine and must get more next week. Am extracting as you told me. Charlie [Uncle Charlie, as usual the know everyone] knows [Carrel] so perhaps that will be a possible way of seeing the T.C. at New York. Streeter, [Director of the Carnegie Institute where Dixon will hold his fellowship] says I may do tissue culture in one of the labs there.

Yes darling next year you will be friend, family, adviser, husband and all. It will be really and truly grand to have just us for a year. I miss you like hell most moments of every day.

[Amélie at Ballycastle Co Antrim to Dixon at Queen's Elms, 12 April 1934]

In Baltimore, Dixon continued his annus mirabilis with a ground-breaking paper on development in the human embryo of the carotid body whose role as a sensor controlling breathing had recently been elucidated. We will come to that later. Amélie's research was less prominent and allowed time for a sybaritic holiday flying round Mexico with Uncle Charlie – many letters! The couple found time to vacation with Dixon's US family and also to make friends in the Carnegie lab. Her year included Baltimore shopping for those back home like family friend and honorary older sister, Elizabeth Kinnaird, who is staying at Lennoxvale – details are irresistible.

My darlingest Amélie, How sweet of you to include me in your price list postcard. I do just feel like treating myself to a few nice things – would you please send me
One Night gown Satin for preference, or in Crepe de Chine in pale pink or white and pink [8/- to 10/- in margin]
4/- one petticote[sic] washed silk in pale pink or yellow
8/- one petticote for best in white or pale pink. Small woman's size
4/- 1 pr stockings chiffon 9" in a good everyday shade (there's a colour not either brown or grey that I find useful, dark)
4/- to 5/- 1 pr Knicks for best, also pale pink or white
It's topping of you to offer and I do hope it won't be too much trouble ... Do say exact cost postage etc I'll send you a cheque.

Tissue paper, rather than tissue culture, seems to have been at the top of Amélie's agenda but she did spend time in the lab and maintained academic plausibility – rather marginally – in publishing one short paper.

4. Coffee at the Carnegie Institute Baltimore – *Amélie on left; Dixon fifth from left. 1934-5.*

ix Amélie and Dixon as Medical Practitioners

Before America, there had been time for a summer of clinical locums. Dixon at Walton Hospital in Liverpool and Amélie as GP in rural County Tyrone where 'they think I'm a Catholic … in spite of my good scotch name of Boyd. What they'd have decided I was if it had been Loewenthal heaven knows'.

'Dr Boyd' for us indicated Mummy's membership of an arcane priesthood. It was also a title, an occasional absence while she was mysteriously 'doing a clinic' and a source of embarrassment when she pressed the cashier at Boots for a medical discount. It is thus somehow a surprise to find evidence of her having been a 'Real Doctor'.

Before deciding to seek a 'house job' at Belfast's Royal Victoria Hospital, Amélie sought advice, longer-term as well as immediate. Medical careers could be a challenge for young women, especially in London. Eddie Bennet, writing from Harley St, had *not the slightest hesitation in advising [you] to put in for the job at the Royal Victoria Hospital ... I have known case after case of women medicos, who have tried and tried and tried again to get [house] jobs in London and have not been successful. ... Rightly or wrongly, there is a definite prejudice against women doctors in London and most of the teaching hospitals have closed their doors upon them. These are the crude facts. If, on the other hand, you have experience in a place like the Royal Victoria, which is very well known, and very favourably known, then appointments are much more gettable! So, apply strongly ... It will give you an excellent opportunity of ... getting over the preliminary stages of any higher examinations you may have in view.* [Dr EA (Eddie) Bennet, 42 Harley Street W.1. to Amélie at Lennoxvale, 27 June 1930, typed]

Walmsley took the same line; his interest understandable toward a student who had achieved a first class in BSc Anatomy, who might get honours in finals and who had, it seems, previously discussed an academic career. Importantly, she was also fiancée (a word they did not use) of a protégé, who was about to join the anatomy department

Dear Miss Loewenthal,

In the first-place congratulations on the passing; and I hope the examination did you justice. I have not yet seen the lists of honours ... Without hesitation, go into the Royal; the backing of a year's work there will carry you anywhere afterwards, and will open lots of doors. ... I do hope you have not given up the idea of academic work. I know you will do well whatever you go in for, but I think you will do very well at teaching and research. However. even for this a 'consolidating year' – to appreciate the problems – will be good ... If you decide on the Royal let me know and I will write to some of the staff. [Prof Thomas Walmsley to Dr Amélie Loewenthal, Lennoxvale, 30 June 1930, handwritten]

After the Royal, where, Amélie told us, house doctors had to pay for the learning opportunity, they both did further hospital work. Amélie was junior to a much-admired, Miss Huxley, at the South London Hospital for Women; challenging but satisfying work which could be busy. The morning colposcopy list

36

goes on till lunch time; this is followed by a harassing 20 minutes in which I write up histories, test urines, listen to chests; in general trying to convince Miss Huxley when she arrives that I am nearly bored waiting for her with everything ready for days. Then we go on till 7 to 8 pm ... exacting and frequent bad moments for me. Last week was fine, F.H. [Miss Huxley] did a Caesar[ian section] for cardiac failure – under spinal [anaesthesia] – the woman was dying; she wanted very badly to have a son, that was why she had risked a pregnancy. Now she is O.K. – or at least getting better and the baby is occupying my other "emergency bed", it is a dear, and a boy. She [F.H.] is so nice, I just felt glad about her for days. F.H. may say anything she likes to me now, I shall never do anything but respect her for the way in which she did that op.

Amélie took London opportunities to extend her knowledge through attending lectures, 'F.H. took me', or through visits to hospitals, such as

an afternoon at Miller General watching C. Joll operate. This was rather impressive, a very fine gastrectomy for a large cancer and a thyroidectomy – apart from the surgery it might have been an "At Home" ... fascinating peroxide blonde with carnation and ether aroma was giving anaesthetics and did the hostess to several GPs and other riff raff assembled. It was a comic mixture of atmospheres. The [operating] theatres there are fine – 2 going at the same time and an assistant who sews up for the mighty and immoral Joll. ['Immoral' referred to gossip about Joll's love life].

Lectures by luminaries of the day were reported to Dixon.

... Langdon Brown [Sir Walter, we will meet his sister, Mrs Keynes, in Cambridge] gave 3 more lectures on Endocrines, they were very good if a bit disconnected, but that is the subject ... Crichton Miller gave us a course on hypnosis which culminated in a practical class in which we had to try on each other ... I managed to hear Keith [Sir Arthur Keith – distinguished, and later controversial, anatomist]. Programme here, in case you'd like to see it; it lasted 1hr1/2.
[Amélie at South London Hospital for Women to Dixon at Greenisland, 16 November and 7 December 1931]

Back in Belfast she wonders about careers. '[G]oing up to town with Dr Allworthy [Eddie Bennet's stepfather] tomorrow to see his skin O[ut] P[atients]. He has been a dear ... I am only just discovering the possibilities in dermatology'.

Dixon was, to us, even less of a real doctor than Amélie. He was Professor, not Dr, Boyd and looked down microscopes in his lab, not at patients. Again, we were wrong. Probably at Norman's suggestion, Dixon had, long before we existed, worked several successive summers at Liverpool's Walton Hospital. This was partly for financial reasons; 'another locum in Walton in September – three weeks. I have already earned £31.0.0 and another three weeks mean £15.15.0, £260 Guineas p.a.' (Amélie got £100 pa at the South London), but also because he valued 'how interested one gets in working with real live patients; the feeling of doing even the slightest tangible good is a relief from academic sterility'.

Dixon told us that he had been house surgeon to Jefferson, pioneer of neurosurgery, in succession to Norman. It is a surprise to us to learn that Jefferson also did general work. 'Norman had an acute appendix and appendicectomy by Geoffrey Jefferson and was for several days very ill'. Richard, driving Dixon to meet, at Cambridge Station, honorary graduand, Sir Geoffrey Jefferson FRS, in the 1960s heard the visitor reminisce that Dixon had been 'my best house surgeon'. Nice, but the letters do not suggest he had worked long for Jefferson and the only reference to Boyd in Jefferson's Manchester University Library archive is correspondence from GW Harris, later Oxford professor, indexed as concerning his failure to gain the Cambridge anatomy chair when Dixon did!

A long letter records that Dixon found the Walton hospital

> *excellent – the largest in the British Isles, 1650 beds and 11 housemen. We have 16 deaths a day and about 1/3 of these are p[ost]m[orteme]d. I have done 5 pms already ... Scurvy, measles with pneumonia, all sorts of carcinomata, phthisis +++, even pellagra have been seen in reality and in the flesh. The common things that are read about but never seen in a general hospital like the Royal. Abortions galore, complete, incomplete and threatening. We get about four abortions a day here. God help them, girls in their teens not knowing what it is all about and*

women in the forties tired with it all, so tired that life, and having children and being married at all have all, long since, lost the zest of being worthwhile. It is terrifying. If ever I had doubts about contraception they have now been dispersed for ever.
[Dixon at Walton Hospital, Liverpool to Amélie at Lennoxvale, 4 July 1932]

The abortions he mentions were those of women admitted with complications. The woman herself or, perhaps a 'back-street abortionist' had tried to end the unwanted pregnancy. Did this letter influence the Harley Street decision a few months later?

He invested free moments in anatomy. 'I only get two afternoons off in the week and, so far, have spent my time going to the Anatomy Dept at the Varsity [Liverpool]. They have a good museum but poor staff'. Subsequently, he is more positive. 'Tudor Jones, the lecturer in Anatomy here, has given me permission to use his Dept on my off days and as he has some very beautiful Lepidosiren [lung fish] material. I hope to make the best of the opportunity'.

Clinical work was very intense. The abortion letter goes on to describe what appears to have been a fairly typical day. It had been

hectic so far. A complete ward round with my chief. Four lumbar punctures and one has to get a manometric reading in each case, two chest operations, and twelve new cases! Then this afternoon I have four p.m.s. It is glorious experience but tiring and no chance to do real reading. I have reached the stage of only reading the medicine of the cases I see and the pathology of the p.m.s I do. I have decided to stay till the 1ˢᵗ August.

He says he was clinically weak but this did not prevent one diagnostic triumph:

... I did spot a cerebello-pontine angle tumour ... a woman in the maternity block with previous vomiting of pregnancy (!). I was called to see her in the middle of the night and found a right sided facial palsy! The next day I went over her C.N.S. and she was absolutely typical. She died today and the p.m. revealed a tumour starting in the auditory meatus and growing backwards into the skull including the

cerebellum. The clinicians, however, think it was not <u>medicine</u> but <u>anatomy</u> – and, in truth, so it was [a very Dixon phraseology].

Collegiality could extend to involving young Dixon in the wider medical world:

... The R.S.O. [resident surgical officer] came into my room at 7.30 a.m and said that he was going to a medical excursion in North Wales ... where I met a number of very interesting Liverpool surgeons and physicians including the renowned Watson Jones who is only 32 and has an international orthopaedic reputation [later Sir Reginald and author of Fractures and Joint Injuries, 1ˢᵗ edition (1940) and 5ᵗʰ edition (1969)] – you may remember his work on fractures of the spine. He was very nice and invited me to visit his clinic at the Royal Infirmary. We had lunch and tea at the hotel there (I shall take <u>you</u> sometime!) and then motored home through Wrexham and Chester. We stopped at Chester to let me see the Cathedral which has fine cloisters and some very good stained glass. Then home to Liverpool at 11.30 after one of the most exhilarating and interesting days I have ever had. [All 4 July letter]

Medicine was sometimes dangerous – no antibiotics then.

... Last Wednesday I felt pretty poor – headache, malaise, dyspepsia and temp. A week previously I had done a p.m. on a typhoid. You can well imagine how I felt about it – and the others here did also. However, the various cultures were sterile and by Sunday I was O.K. They are taking a Widal [typhoid test] tomorrow just to make sure but it will be O.K. I know. [Dixon at Walton Hospital Liverpool to Amélie at Lennoxvale, 25 July 1932]

Both parents did locums in general practice; and found them lacking. Amélie writes that the practice's owner

showed me the surgery and there's practically nothing but sod[ium] bic[arbonate] and Chloroform – nothing to test urines – he 'never needed that' – no 'etc.' for emergencies – serum, insulin or the like, those are quite beyond the pale; and all he would tell me was – charge so and so, so much and that means another 5/- till I could have been sick. It's awful. They're all poor farming people here, some insured but with those it's the same old story, [illegible word] but no

examination ... They need a real doctor frightfully badly – and the next one near here is 74 and deaf and blind ... I got back 6-ish and then began a fine evening! – A man waiting for me to go 3 miles to see his 'Aunty'. I got there up precipitous mountain lanes and 'aunty' was in uraemis [?] coma and had been for 24 hours. She was 84! Hospital no sense and they didn't want it either so I gave her MgSulph P[er]R[ectum] and Coramine wh[ich] I'd got from Aileen [former fellow medical student and later wartime colleague] but of course it didn't help. Intravenous saline – or even P.R. – a remote phantom as there wasn't even a decent needle let alone NaCl [saline solution] in the surgery. The only good thing is the people are so sane about death. You'd have thought it was the weather we were talking of. They were much more curious and interested in me than in anything else. So I've left her to die. I've to dispense everything, and no concentrated solutions, but so far, no troubles that I can see; the surgery is like a back yard. No light, no hot water, no heating of any sort to boil water on. I bought a spirit lamp and rigged up a stand with wire and a saucepan! Midder [midwifery – childbirth] any moment I expect. I also got stuck in a lane in the car last night 3 men came and pulled and lifted me out but I wasn't back till 11.30. The country is <u>lovely</u>, beautiful lights[?]. I'm staying in the 'Commercial hotel' – great fun. I have meals with 'commercials and Fathers' and am seeing life. Altogether I love this game. Probably more still in retrospect, as anxiety rather spoils the view for me at present.
[Amélie at Commercial Hotel, Pomeroy, Co. Tyrone to Dixon at Walton Hospital, 9 August 1934]

Dixon's locum general practice, in Balham, near Amélie's South London Hospital, had a similar money-making emphasis (and no telephone) but at least it was close to her. He introduces his arrival with typical emphasis.

Dearest Dearest Amélie, Is it true, only two more days and I may see you? Certainly before the week end ... I am in a poor class general practice which nets its possessor about £2600 a year. (Incidentally he pays me £8-8-0 a week. (8 x 52=420 ie £2200 a year profit for him but that's by the way) ... I have a surgery from 9.30am-11.30am each day and from 6pm -9pm each evening! Some visits between 11.30am and 1pm. More or less free during the afternoons, absolutely free after 9pm except for midder cases – (about 3 are due). Do nothing at all on Thursday afternoon and on Sunday the whole days. ... So, if you can arrange anything about your off time, I might see you! My

phone number (private house not surgery the surgery does not possess one) is 0081
[Dixon in Balham, London to Amélie, 2 September 1931]

After that foray into general practice, Dixon's clinical career was over, save for a brief wartime postscript operating with Norman.

Amélie worked to achieve a postgraduate Diploma in Public Health which award gave some validity to her irritating refrain: 'it's my medical training' used when criticising allegedly unhygienic housekeeping by daughters-in-law but which did not lead to a career. Part-time work in general practice continued until the end of the war. Thereafter her medicine was restricted, like that of her friend Biddy Barcroft and many other medical wives, to 'infant welfare clinics' and 'school medicals'. Such roles were mocked as only providing 'pin money' rather than being 'real medicine'. Once we had left home, she, now in her 60s, enjoyed an expansion of her work into 'contraception and family planning'. 'Marital dissatisfaction always starts in the bedroom' she would comment. As school doctor at Impington Village College 'where I do the medical each week' she was proud of developing, jointly with its headmaster, an innovative approach to what was then called 'sex education'. Not the career Bennet or Walmsley had expected but a career nevertheless and, for her, the earlier clinical years had also had more significance than we realized.

Darlingest Dixon, I don't believe I would have "waited" even for you if it had been me with only one life! That's where I am so damned lucky, with the chance of having had jobs like R.V.H. and this and interest and people ... economically of some (very slight in reality I know) value, but which makes a great difference in feeling an independent person with a Right [sic] to do what one wants.
[Amélie at South London Hospital for Women to Dixon at Greenisland, 18 March 1932]

The medical discount in Boots the Chemists symbolised something more than pennies.

As the medical daughter, Amélie had greater involvement in her mother's case than the other sisters. When Elsa first 'broke down', cousin Eddie Bennet was consulted and she was sent to

Switzerland to Pranguins, near Geneva. Pranguins was *the* institution for the well-connected mentally ill. A letter from Amélie's sister, Helen, confirms that; the volume she mentions is still in print.

> *[H]ave you got or can you get hold of 'Tender is the Night'? It's by Scot Fitzgerald (American) husband of a woman I met at Pranguins (and all about P.) There is a grand description of the place, of Forel, and everyone there. It's terribly well written but really horrifying how the man uses his wife to provide copy. But I could not stop reading it.*
> [Helen Loewenthal at Courtauld Institute, London to Amélie in Baltimore, 7? October 1934]

Forel was the Pranguin's director and, naturally, a friend of know-everyone Uncle Charlie who informed Amélie that when he 'lunched at Dr Forel the other day ... they both enquired for Dixon [presumably introduced during his summer at the League]. Both liked him very much'. It is not clear whether 'they' refers to mention earlier in the letter of pianist Horowitz or to other Forels. Father and son were both renowned psychiatrists; the father's posthumous reputation later blemished by accusations of racism. Letters from JMcC show the younger Forel to be eclectic (and expensive) in his choice of treatments. They ranged from, in today's parlance, family therapy:

> *He proceeded to talk about Helen and thought he had done her a great deal of good ... if he is right, and likely enough he is, the labourer is worthy of his fcs. [Swiss francs] 300 ... he would like to see Peggy ... I think he really takes a genuine interest in the family ... he expressed a very high opinion of you and said you were "ein Pracht-Kerl" - "a splendid fellow".*
> [JMcC at Hotel Beau-Rivage, Nyon, near Geneva to Amélie, 30? January 1932]

through supervised outings from the clinic:

> *I suggested allowing M[other] to join me for the week-end at the Beau Rivage which she was very keen to do. He agreed with the week-end idea but preferred Divonne. He doesn't like his patients to stay at Nyon – too near and too much gossip. It's all the same to me and Divonne will be more of a change. So, we will go in the car on*

*Saturday forenoon and return Monday forenoon. M's old Pflegerin
[nurse] will – by Forel's wish – go incog[nito] by train though I don't
anticipate for a moment that I shall need her.*

to (unsuccessfully) the latest pharmaceuticals:

*He attributes M's improvement more to my influence than to the new
gland medicine. Says he has obtained excellent results with it in men
patients, but little or none in women.*

Maybe conscious of Helen and Eddie's entwinement, JMcC
spurns advice from Helen who is

*recommending something in the nature of psycho-analysis ... I have
never taken any notice in my replies to her. To discuss it with her
would serve no useful purpose. But you are a doctor and accustomed
to face things as they come.*
[JMcC at Hotel Beau-Rivage, Nyon to Amélie, 10 February 1932]

Perhaps so, but tough to be *ein Pracht-Kerl* in dealing with a
mother.

Elsa returned home from Pranguins, but not for long. By
late 1932 she was resident in an Edinburgh institution, Vogrie
House. She remained at Vogrie until her death a generation later.
Of that long institutionalisation, financial, but not clinical, records
remain. 'Residence' in 1933 costs eight guineas weekly and
nursing three guineas. The cost, some £700 a year, was remarkably
stable. By 1947, in the last year before the NHS, and the last for
which Vogrie records can be found, the rate for board and for
nursing had not changed. The same weekly fee is charged as an
extra when JMcC or Amélie ('Dr A Loewenthal') stay to visit.
Amélie is charged four times in 1932-3 but none of the other
daughters are. Elsa is, from time to time, briefly in Ireland to be
with JMcC. During those occasions, the cost of being
accompanied by her Vogrie nurse is also charged.

As Elsa ages, should she have new treatments coming
into clinical practice; leucotomy for example? Amélie takes
advice and JMcC responds. It is the final month of the war in
Europe. Let him have a last word on his wife, our grandmother.

44

I feel that you did very right to have a talk with Dr Slater [Eliot Slater, 1904-1983, leading psychiatrist at Queen's Square/Maudsley]. Quite apart from the desirability of keeping yourself abreast of modern research and advance in medical practice, I realise and share your earnest wish to explore any possibility of a happier solution for M[other]. But I cannot bring myself to approve of the suggestion of Dr S[later] ... This reluctance on my part is not due, I think, to the apathy or indolence of old age, but to more cogent reasons. Apart from the obvious ones of age – M was 70 or 71 last year – length of illness, and waning physical strength (Helen wrote she was more quickly tired), there are other factors, among them family history. I won't go into that now, though I am quite willing to discuss it frankly with you or Dixon when opportunity serves [referring perhaps to the suicide of Elsa's elder brother and of cousins or to alleged syphilis suffered by Elsa's father]. In that history there is nothing, I am convinced, to cause any of you the slightest uneasiness or misgiving for yourselves or your children. But taken together with all the other circumstances of M's case it is a ponderable.

Though I have no claim to medical knowledge I tend strongly to the belief that mental illnesses is, in the main, due to physical causes. It is likely that some day physical methods of treatment and cure will be found. Things seem to be tending that way. But that time is not yet. These modern methods of shock treatment and surgery are still in the experimental stage – empirical as you say – and the 'why' or working of their effects not understood. I feel more than loth[sic] to risk the experiment for M[other].

Prof. H [Sir David Henderson, Elsa's Edinburgh psychiatrist] may be 'conservative' in the view of younger research workers. He always impressed me as being outstandingly 'sane' and at the same time open-minded. That he is not 'stick in the mud' is proved by the fact that when you and I together spoke to him some years ago he was already using the shock therapy in cases deemed suitable. Dr S. himself told you, you say, that Prof. H. was 'much in favour of leucotomy', so it is clear he is quite aware of what is being done and has probably tried it ...

I have tried to write you what I feel about it. It has not been easy, and I am sorry to bother you with it just at this time, when [mother of three, 36 weeks pregnant with fourth and exiled from half-destroyed home] Lord knows, you have enough to think about and do. If you yourself feel like asking Prof. H. about it, and if Dixon approves of your doing so, I of course offer no objection. As an interested fellow-medical you have a better plea.

As you say M. has good care at V[ogrie] and is not unhappy.
My hope was, for a good while, that it might be possible to bring her
home and provide proper trained care. But that hope too is waning
as I feel increasingly the natural effects of my age [81].
    War news thrilling. Surely the slaughter and destruction will
now end soon.
[JMcC at Lennoxvale to Amélie at 12 Stormont Rd London N.6,
2 April 1945]

We children saw Elsa once only, at Vogrie. She took little notice of us. As with the other grandmother, only mementoes substitute; initials on the suitcase that held the letters, and paintings which hang in our houses chosen by her in earlier years. Her only grandmotherly role was indirect, Amélie's would return with, special treat, Edinburgh rock, from her six-weekly trips to visit.

Elsa was still in Vogrie when she eventually died in 1959. The leucotomy surgery she didn't have now has a bad reputation. By curious coincidence, the second wife that JMcC's cousin married, after his first, Elsa's sister, died, also became seriously mentally ill. After her leucotomy, she improved sufficiently to run, in widowhood, the London branch of Iklé Freres.

2 Cambridge on the Margins of War

i   Getting Established in Cambridge

By September 1935, back from the USA and the Carnegie, Dixon is in their first Cambridge home, 9 Newnham Terrace, just round the corner from Gwen's grander house. Cambridge was thenceforward, despite two London intermissions – one brief, one longer – their place. It was to remain the focus for more than half a century until widowed Amélie sold their final Cambridge home.

Dixon and Amélie wrote only when apart so letters cluster round separations. Over the years, until the death of JMcC in 1952, a cluster usually means Amélie is in Northern Ireland. That autumn was one such period. While Dixon was settling into Newnham Terrace and into his role as demonstrator in anatomy, Amélie was moving towards a different life. In the final trimester of pregnancy with John, she is supported at Lennoxvale. 'Daddy made me breakfast in bed Sunday a.m., came up complete with rose from the garden to see me'. "Made" doubtless meant suggested to "the maids". Margaret would have cooked, and Sarah carried. The expected infant, with sobriquet fitting either gender, is reported as vigorous. 'Fish kick[ed] all over one side again so fish only one I think'; singleton status a relief as both grandmothers had been twins.

By Christmas, Amélie had rejoined Dixon, and they had moved into 11 Bentley Road, a substantial detached Edwardian house. Its ownership is not clear: bank loan, mortgage, paid for by JMcC, or rented? The latter seems likeliest as there is record of dispute over whom to blame for water damage from a leaking roof. Bentley Road seems a very substantial residence for a junior academic. Census records tell us that six other households, generally more senior and more distinguished, preceded or followed the Boyds into that house. Five of the six, husbands were fellows of the Royal Society; two, nominees for a Nobel prize.

Reunited in Cambridge, Fish became John in January 1936, at the Brunswick Nursing Home.

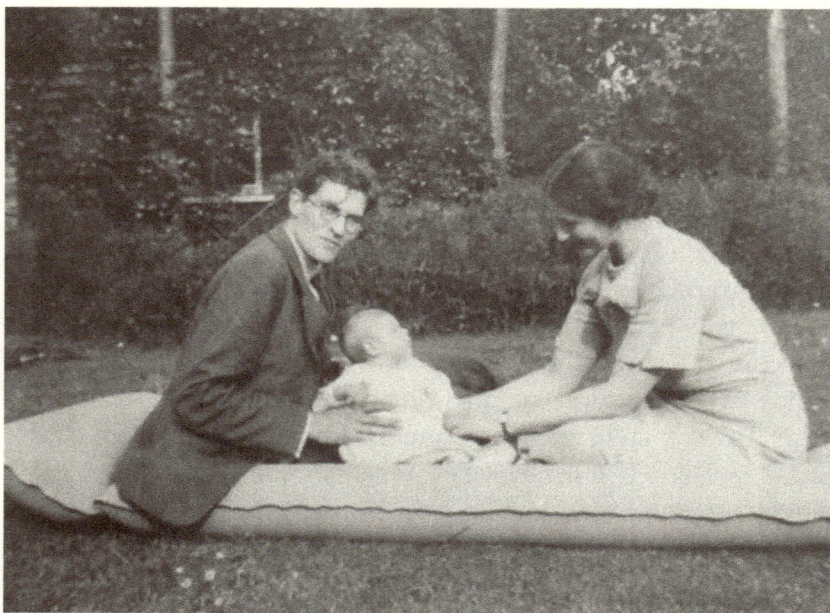

5. Parenthood – *Amélie and Dixon with John, 1936.*

Dr Amélie Boyd became Mummy. Her and Dixon's roles diverged accordingly and according to the times. Motherhood certainly spelled the end of Amélie's academic aspirations. Her detailed interest in Dixon's research also waned though vigorous social involvement in his professional life did not. Wives were generally excluded from formal college and university activities but that was absolutely not the case in the domestic setting. Life in Bentley Road soon became expansive with staff (nanny, maid and gardener), dinner parties and overnight visitors. Amélie threw herself into that life. Dixon's colleagues came to meals, Belfast and transatlantic figures came to stay. Creating and conducting a busy and hyper-hospitable domestic scene was absolutely Amélie. It continued throughout the marriage.

Dixon's career was on a more than satisfactory path. A teaching fellowship at Clare College attracted letters of congratulation. 'How delighted we all are, here at Queens (and down Lennoxvale) at your election to a Fellowship' writes QUB vice-chancellor with a nod to JMcC. Dixon remained a fellow, with interruptions, for the rest of his life.

By July the academic year was over. Amélie, and John, and Nanny, were again at Lennoxvale and the letters flow. Dixon has moved into college for the *vacation* despite maid and gardener at Bentley Road. Amélie orchestrates use of the house by friends and by Dixon's stepfamily. She also orchestrates, again, Dixon's physical wellbeing – 'You'll have to go to a dentist sometime, foolish to let them go and no economy – so there says your wife'. Following report of a headache, she enquires: 'is it your glasses do you think? If you think so, do you think you could coax yourself to go and make an appointment … or will you wait till I bully you?'.

The college fellowship was small in number, friendly, and of course entirely male. Its members, mostly from grammar schools, said Dixon, were or became academically distinguished (distinction remains a key mark of fellowship but there are more than ten times as many; the current 'master' female). The younger fellows were especially convivial. With Amélie in Belfast, a relaxation for Dixon could be

> London with Hammond – leaving at 6 in his car (black tie, I wonder can I still tie it!). We go to Ballet with Spooner, Greaves and his fiancée. I am looking forward to it
> [Probably July 1936]

Those four were to become lifelong friends. Other amusements are mentioned in the same letter. 'To-night the College entertained the servants. Dinner and Concert. They were frightfully pleased and very un-self-conscious. the "lower classes" are charming (*pace* the shades of your Auntie Olga!)'. A bit rich coming from the Greenisland boy.

Dixon reports with satisfaction the 'prospect of a bunch of papers – if I ever write them'; slowness in writing up research becomes a recurrent theme. Lecturing to large audiences scared him and he later told us of vomiting with anxiety as he walked over Silver Street bridge on the way from Newnham Terrace past Gwen Raverat's house.

Domestically, matters were well-managed by Amélie, but "the department" soon became a problem.

> *Work is going well at moment but HA [Harris] is worrying me a lot. He and JS B[axter] [fellow Walmsley product] are thick as thieves and yesterday HA took me aside and said that there might be a lectureship and if there is he thought I should not get it as I hold a Fellowship. On the surface a good enough excuse but why the hell could he not have said it when I got the Fellowship. A lectureship should not be on basis of income but of ability to carry out the duties and a new man does NOT know the ropes. However, he can do what he likes. All I do know is that I will not toady to him. He is a most obnoxious person and has not even the knowledge of anatomy that would excuse his obnoxiousness. I am going to speak to Thirkill [Master of Clare] about this and sundry other incidents and get his advice.*
> [Dixon at Clare College to Amélie, probably July 1936]

Thirkill notwithstanding, or perhaps at his suggestion, Dixon decided to leave. According to her later account, Amélie was recovering, in the Brunswick Nursing Home (from being delivered by the GP, forceps and all, of second boy, Robert) when Dixon informed her of his 1938 appointment to become University of London Professor of Anatomy at the London Hospital Medical College in Whitechapel, East London. (Walmsley had been right. Cambridge could then, as now, be a good stepping stone). Dixon was 30, early for a medical scientist to get a chair. He thought she would be delighted; allegedly, she wept. It was expected to become a permanent move away from Cambridge.

Finding a suitable London house seems to have been a challenge. It is not until ten months later that, their move now imminent, Dixon pens in characteristic style a farewell to their Cambridge home. He is staying in "digs". She is coping with the move.

> *28th March 1939*
> *St Philips Vicarage, London E1*
> *Monday Night.*
> *Darlingest Wife,*
> *Just a note – Bentley Road has been so pleasant. Three and a half years of real married life without the strain of newness that leaves a SWEET*

SWEET bitter tang in my memories of Queens Elms, and without all the distraction and 'homelessness' of America in our 'Wanderjahr'. It has been so pleasant, so very very pleasant. Cambridge, Home and Garden, Clare! The boys [only two then] – it will always be for us their birthplace. All our intimacy and snuggling down before fire in study, or on Veranda or in a double single bed! Really something to be proud of – to have found our feet so sturdily and promptly in an alien land and among the cold English. ... I am more in love now than ever before ... Bentley Road or Timbucktoo – it's all the same. So, keep that in mind in the inevitable depression of the next week.

'Alien land' or 'cold English' were not overt musings of the Dixon we knew. He was, for us, completely English, and, of England, Cambridge. Ireland was spoken of, if at all, as a distanced memory.

ii   London Briefly

Fernside, Theydon Bois, on the outskirts of Epping Forest, had been eventually selected. It was well placed for Dixon's new workplace, with plenty of room for children. Robert pulled himself to a stand in a cot there that last summer of peace, an early memory. Gardener and housekeeper were recruited and vegetables planted for the new life. Nanny moved with us.

It was all in vain, and brief. Six months later the German Army was in Poland, Amélie and the boys were in Ireland. Within the first week of war, a harassed evacuation organiser reported on her use of Fernside. It would be paid for.

200 expected children did not arrive, and 59 expectant mothers that had not been catered for did ... I managed to find accommodation for ... eight at Fernside. Mrs Gillies helped to shade windows with brown paper etc and has bought less powerful bulbs, all yours have been put away, also the things in the nursery, cot etc have been stored in the dining room and that door locked, so they have the use of five bedrooms, servants sitting room and kitchen ... I am enclosing the forms for you to fill in [for] the Post Office [to] draw £2 for the first week due to you ... your gardener was very worried because the rabbit has disappeared – [pet for us, not food]
[Kathleen Bray, Westwood, Theydon Bois to Amélie 9, September 1939]

Medical school administration was even more proactive. Nine days before war was declared Dixon had sent Amélie a "wire" – no punctuation in wires.

> *AM IN LONDON TONIGHT AS HAVE BEEN APPOINTED DEPUTY DEAN PRO TEM STATE OF STRAIN DECLARED AND PRELIMINARY EVACUATION OF HOSPITAL MAY GET TO THEYDON TOMORROW LONDON QUITE NORMAL LOVE TO YOU ALL = DIXON*

He is in Cambridge within the week 'making some arrangements re the London Hospital'. Dixon, though London university professor, would return to Cambridge for another three years. His evacuated department moved into the geology building. His fellowship at Clare was renewed.

> *It is a great joy to us all, and if I may say so to me personally that the fortune of war has made it possible to keep you with us for at least another year, and I am sure that if you are still in Cambridge when the next Michaelmas term comes, it will be the earnest wish of us all to have the privilege of re-electing you again.*
> [H Thirkill, Master's Lodge, Clare College to Dixon, 30 Barton Rd Cambridge, 21 October 1941, handwritten]

Amélie had been on holiday in Ireland when crisis struck and stayed on because, as she writes a few weeks later, 'safe escape from an air-raided England would be very hard'.

iii Separated by the Irish Sea

Several things strike us on reading the, sometimes daily, letters during what became a long separation. Most notable is the change of tone after May 1940 with the fall of France and the start of bombing. Initially, war at a distance, Dixon is organising teaching and examining and, in research, 'getting some work done but it's very hard to concentrate in these times'. Amélie, John and Robert have moved from Belfast – chauffeured by JMcC's Alec – to join Aileen, the medical student friend, in her Armagh general practice.

*Nanny came along ... The work here is slack. But it is work. I have
the use of a car ... I think we will work out finances at £5 per week
each way – for billeting us – and for my salary.*
[Amélie at Gilford, Co Armagh to Dixon, c/o Wooster 339 Cherry
Hinton Rd Cambridge, 2 September 1939]

Practicalities predominate.

*Car registration is not astray but in a large manilla envelope and, if
not in the car, in the centre of my desk [at Fernside] with other papers
investments etc. I feel sure it should be with these. And if you have the
papers that you collected on the first visit to the house you probably
have that envelope there also. But don't worry – Alec suggests getting
the last three months refunded on Tax and Insurance.*
[Amélie to Dixon, undated]

The early ebb and flow of war news only rarely elicits comment.
It is mostly ebb but earlier, within a month of invasion of Poland,
Amélie is optimistic of reunion. Despite the Russo-German non-
aggression pact between Molotov and Ribbentrop, she hopes

*if air raids don't result this week as reaction from present diplomacy
in Russia etc we might re-consider coming together again* en-famille.
[31 October 1939]

But they don't, not for two years, not until the autumn of 1941.
Teaching medical students was a "reserved occupation"
so Dixon was not "called up", a later mild embarrassment to the
older boys. The Boyds' war was to be at the margins of the
fighting. For Robert, war was normality, as was sharing everything
with John who, being older, knew about things and could whistle
and ride a two-wheeler while he was still on a 'trike'. Uniforms,
aeroplanes, guns on the Belfast ferry were just how things were.
The grown-ups' "before the war" was as mythical as the tales
Mummy told of 'when I was a little girl'.
After John's fifth birthday, the family were reunited in
Cambridge, living first with the Godwins on Barton Road, he
another Clare College fellow. Then we moved round the corner to
Grantchester Road, number 26. Italian prisoners farm-worked the

meadow across the road. We chatted with them about their children. Wrigley's spearmint from uniformed American airmen was normality. So was collecting rosehips to support the war effort; for vitamin C, we were told. That was in the garden at Miss Cooke's, the Eton of Cambridge nursery schools to which on the child seat of Amélie's bicycle (Robert) or pedalling his two-wheeler (John), we were taken each morning (the nursery occupied the ground floor of Professor Cooke's substantial house. He occasionally come downstairs to grace us with his presence). The man-eating tigers at Whipsnade, seen with a frisson of excitement on a later birthday outing, were assumed, at least by Robert, to be fed on Germans – "Jerries". It took him three decades for ingrained hatred of that nationality to fade to imperceptibility.

But, for the 23 months before the reunion of 1941, Amélie and Dixon's lives were separate. For them, normality it certainly was not. Neither had a fixed abode; he in Cambridge, she in Ireland. Their letters indeed flow, 132 survive. The reliability and regularity of wartime post astonishes.

Life in Ireland was unsettled. In a complaining tone, unusual for her, she tells Dixon that

> *I am just worn from balance ... by the rootless concatenation of moves, changes and anxieties since Robert was born – starting then with my return to a maidless house! It has been a nightmare with lucid intervals of a few weeks ... the five re-adaptations of myself and the children since the war started ... have become a job, cots – food – packing – etc.*
> [Amélie, Osborne Park, Belfast to Dixon at 339 Cherry Hinton Rd Cambridge, 14 January 1940]

More usually she is positive.

> *I have scarcely unpacked yet and still feel like a cat in a new house – but you know me – and my adaptability once I see 'people' again. And someday all this 'shifting' will be over – and all we'll want is peace together and lots more children.*
> [Amélie, 16 Malone Park Belfast to Dixon at 30 Barton Rd Cambridge, 17 January 1941]

The letters between them those months are to and from at least six Irish addresses. Two are in Newcastle where at one, Streamville, childcare duties were shared with Biddy Barcroft (fellow "woman doctor", wife of Queen's professor of physiology and lifelong friend to be). A third address was the Barcroft's Belfast home in Malone Park, while Aileen's general practice, rented rooms in Osborne Park and brief stays at Lennoxvale are others.

Dixon, in Cambridge, boarded with the Wooster family at 339 Cherry Hinton Road – Peter is Communist (according to his daughter, not a card-carrying one), a crystallographer and fellow of Peterhouse, Nora has Belfast connections. Then, Dixon was with the Godwins at 30 Barton Road. Clare-fellow Harry is a distinguished botanist.

Both families offer to provide space for three more Boyds should Amélie, and John and Robert, return. Margaret Godwin, in a typical example of the mutual support between friends apparent in many of the letters (and of continuing class assumptions) offers

> *3 vacant bedrooms or 3 ½ if you would not mind John sleeping in David's room – this may be more accommodation than Nora [Wooster] has free just now ... I have to change my maid anyway at the end of the month and if you decide to come, I will get a daily maid. I suggest you talk this over with Dixon and Nora and Peter [Wooster] and see what is best. It would be a good idea if your nurse would come too: try to arrange that if you can.*
> [Margaret Godwin, 30 Barton Road Cambridge, to Amélie, 7 March 1940]

Eighteen months later, her offer was accepted (Margaret, in Harry's Wikipedia entry, is noted as being an innovative botanist. To us she was just another mother and rather a conventional one). Nanny rejoined us, and the flood of wartime letters dried to an occasional trickle.

During those nearly two years of separation, punctuated by occasional joint holidays or brief meetings, love, happy memories, and delight in "nature", and in the children are recurrent themes. So are fantasies.

Worked hard until 4.10p.m. and then cycled to Grantchester by way of Trumpington. As soon as I reached the Botanic Gardens you were with me and we went out past the end of Bentley Road, under all the Trumpington Road trees, past the large field (on the left just before the village) of yellow butter cups and off to the right past ... the little church with the brass. There I stopped and talked to you about the tulips which of course reminded me of John's liking of them and our conversation then became very intimate, both knowing what the other said. Then off I started again down the hill and across the bridge opposite the vicarage and we had our usual laugh.
[Dixon at the Godwins *'as from Clare College'* to Amélie at Streamville, 28 May 1941. *'EXAMINER' 2703*]

Note the examining censor. Occasional letters have sections snipped out.

One Amélie-reply is typical of many.

My very darling, how I wish you were with us here! Sometimes, as this afternoon, I catch a mood of delight because of the children. We are sitting on the pebbles, by the bridge where we once sailed orange peel boats with John! It's warm and sunny – John is hopefully trying to catch trout by paddling in after them Robert helping him by shouts and stones and in between drinking the river with a tea spoon.
[Amélie at Streamville, Newcastle, Co Down to Dixon at Godwins, 30 June 1940]

Dixon "fire-watches" for bombs in rotation with Harry Godwin. 'Last night was quieter and I had a good night's rest – the first for five nights'. He is (letter 'Opened by examiner 1309') 'relieved when it rains – thinking – they won't come tonight'. Oxford's Le Gros Clark does more, 'spending his summer holiday as a stretcher bearer in London! Makes me a little ashamed!'. It is surprising to learn, in a nice note from the Clare Bursar, that fire-watching is not only paid but that anxious nights count as taxable employment.

Dear Boyd, ... I am sending you a cheque. The College has every reason to be grateful to you and no cheque has ever given me greater pleasure to sign. I shall in due course inform the inspector of

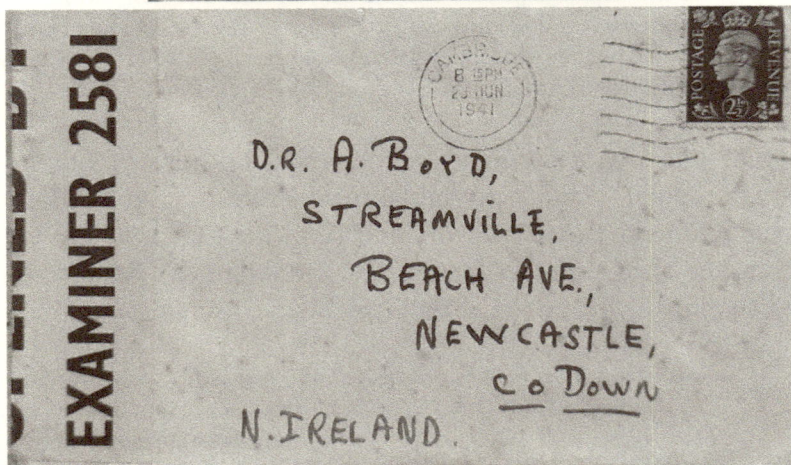

6. Wartime – *a. Permit for travel to Northern Ireland.*
b. Censored domestic letter.

taxes that payment for services rendered has been made to you retrospectively as follows: year to Oct 5 1941 £25, 1942 £50.
[L.J. Harrison to Dixon, 19 October 1942]

Grass-widowed by Amélie's absence in Ireland, Dixon was invited by Norman and Alice to have a break from nightly fire-watching. The night was spent, not in sleep, but in assisting Norman operate on victims of Norwich bombing. Amélie writes she is 'Glad you are safely back from your night of Blitz'. Dates don't tally for Norwich's biggest Blitz; it must have been another raid. (Amidst the mayhem, bibliophile Dixon looked into the bombed library of the hospital, the equivalent of a bombed bar to an alcoholic. Perhaps he succumbed).

Generally, academic life remained remarkably unchanged. Two months before Dunkirk and as British forces are driven out of Norway, Dixon reports that

> *last night I went at a lecture with Peter [Wooster], on geology of South Africa. Intriguing very early pre-Xtian gold mines and beautiful pre historic ruins of small oblong stone in all sorts of patterns [small sketch] chevron and herringbone chiefly. Much in advance of Irish forts ... [and a few days later] ... I do love you – but have no time to write! Just finishing ours and the Bart's M.B.s and now have got the Oxford papers. And my own Clare exams to-morrow and next day.*
> [Dixon, London Hospital Medical College at St Catherine's College Cambridge to Amélie, Belfast, 5 and 13 March 1940]

A year later, with Germany across most of Europe, his days remain regular. 'I come to the lab in the morning, go to Barton Road [Godwins] or Clare for lunch and then back to lab until seven and then Clare or Barton Road for dinner'. Scholarship, and supporting the young in their research, continue too.

> *I am in the lab [he writes on a day Belfast is bombed] ... I did the pictures for the book [to become Hamilton, Boyd and Mossman, Human Embryology, ed. 1 1945] and am now relaxing before starting on a paper for the J[ournal] of Anat[omy]. Angus [clinical student about to qualify, who we will hear more of] is here with me. He came for the week-end to do some work on my viper material – Jacobsens's organ. Yesterday, too,*

I had Shillingford who is starting something on the Quellzellen in the kidney. It is work I once meant to do myself so I am very interested in his findings He is a good lad and very hardworking and will, I think, go quite far. [he did; became a leading British cardiologist]
[Dixon at Godwins to Amélie in Newcastle, 1 June 1941]

Shillingford crossed the Atlantic to finish his medical degree at Harvard, funded by Rockefeller. He updates gratefully, and includes a methodologically-innovative suggestion for future research.

Dear Sir ... during the passage through the rather high pressure system it was difficult to keep the true scientific approach which you, above all others, taught me ... I cannot remember whether I mentioned in previous letters that after consulting Bodian at Baltimore we managed to get his technique [of nerve microscopy] [to] work very well on the kidney. The result is most interesting as nearly all the [nerve] fibres appear to be going to (or coming from?) the cells of the juxta-glomerular apparatus (or rather the Quellzellen in the wall of the afferent arteriole) ... It would seem that the next step might be to prepare a living animal with recording electrodes on the nerve supply to the kidney (taken off the renal vessel) and then perfuse the kidney with various agents and record (if any) variations in potential.
[JP Shillingford at Presbyterian Hospital New York to Dixon, 28 December 1943]

At a college dinner hosted by the master, Dixon

had a long talk with Mrs Hutton [née Schuster; her father, German-born Sir Arthur Schuster]. She is magnificent, so logical and cool about the war. And very illuminating about the education of her children. She said she felt it had been such a pity that she had given a completely international attitude to her son, who now is unable to see that anything good can come out of the war and who does not much mind what happens. He has, she said, no patriotism. Isn't it extraordinary that we have reached such a stage in the world when a Mrs Hutton has to be sorry for having done the right thing!
[Dixon at Godwins to Amélie, 2 May 1941]

For Amélie, life has been very different. Two months after the German invasion of Poland she can still

foster a tiny spot of hope for the relatively rapid collapse of this war on one hand, or, failing this a mild chronicity which would allow of our return [to Fernside] In the meantime 2/6 p[er] week to Mrs Maynard keeps it and we are fairly solvent even with the increased I[ncome] Tax.
[Amélie at Gilford General Practice to Dixon, 27 October 1939]

At that stage, day-to-day life in Ireland remains pretty normal. Judith Iklé, teenage Swiss cousin having – anachronistic concept – a gap-year working in a Belfast nursery, stays on till Christmas supporting Amélie and helping with the children. Her father, Fritz, writes from his home, a lake's width from Germany, that

I do not think she could be in a better spot ... We felt very uncomfortable for a few weeks here, but today it looks calmer.
[Fritz Iklé St Gall to Amélie Belfast, 31 October 1939]

Switzerland continued at peace throughout, but he was wrong about Belfast. It was not to be a 'better spot'.

By May 1940 there is a change of tone. Judith has gone back to Switzerland; Norway is lost and Chamberlain has resigned. Letters are daily. Churchill becomes prime minister and Amélie's view is similar to the new prime minister's more public assessment.

The war is really on us now. I wonder if we can pull ourselves together in time to win and we must. But will we? The German attack seems to be swaying back and forward. At least one feels that this acute struggle will end either way more quickly than siege ... If the Germans beat the Dutch as they seem to be doing, they will need to consolidate their position on the continent before attacking England. That means many weeks yet. Then will it be invasion or areal [sic] attack of ports and munitions? Daddy thinks the latter because the navy would make an invasion difficult and in that case parachute troops useless ... I feel relief that we are fighting at last and was in much excitement this morning and felt myself capable of great ferocity ... How grave the situation is darling ... I feel the situation changes so rapidly that by the time my letter reaches you some other dreadful event has already occurred.
[Amélie at Lennoxvale to Dixon at Godwins, letters of 12, 14, and 15 May 1940]

Mundane health worries about the children are shared. 'I am with your Robert just now ... his cold is better. Frank saw John y'day and said there was nothing in his chest. I phoned Wheeler who will take out his Ts and As [tonsils and adenoids] on Friday 24' [They were].

France falls. Assets need to be portable and plans flexible.

I today got £50 in £10 notes and have them in Belfast ... the pearls are with me here! If Ireland is invaded as a means of attacking England from behind you must remember that I have always my medical degree to barter ... Joan's cottage, by the way, is above Bloody Bridge and would be a possible haven sh[oul]d you ever find we were not at Streamville!

My dearest dearest love – keep well.

[Amélie at Lennoxvale to Dixon at Godwins, 25 June 1940]

Earlier she had thought of rejoining Dixon without John and Robert but was [27 October 1939] 'unable to make headway with Daddy ... He won't let me leave them in Lennoxvale. I can't help reflecting on what mother's reaction would have been both in pleasure at having them and understanding of my absolute need to be near you'; a poignant comment in the context of her mother's then-state. We were told that Amélie's sisters had felt our being left at Lennoxvale might have upset the servants so persuaded "Daddy", JMcC, to refuse. Amélie instead rented rooms in Osborne Park. '6gns per week with a nurse ... £1' to look after us. Very occasionally over the years ahead, a hint of resentment for this rejection directed towards a sister, not the father, might resurface.

Comments on the wider scene intersperse the personal. Britain bombards the French navy lest it might support German invasion. 'Atlee spoke well to-night. The fleet attack was tragic and yet comforting'. When Mussolini chose Hitler, she feels

sad about Italy. I wonder if John and Robert will see Florence and Ravenna ... the Jugo-slavs [sic] are fighting which is more than any of us dared to hope for a week or so ago. If they can put up anything like the show the Greeks have, it will completely change the future of

the war ... the Russian attitude is obviously hardening and may well give us some further reason for hope.
[Amélie at Streamville to Dixon at Godwins, 12 June[?] 1940]

The hope was eventually to be justified but, ally-to-be or not, Amélie had already, by the start of the war, given up on Soviet Communism as an ideal for the future.

> *I bought some chiffon underwear and a nightie for myself. I hope you'll like them! My last fling in a capitalist world ... If Russian technique of propaganda in Russia before and during their war [probably referring to September Soviet invasion of Finland]; of aggression in spite of protestations of peace; of calling a spade a violet; of disbelief of the individual's capacity to reason: if this is the full stature of socialism we need not waste our time fighting for that. Then, what we have; injustice of opportunity, false standards of freedom, capitalist oppression of the workers is no worse, and all our hope lies in the individual's strange and occasional desire for the evolution of a wiser state through education, and gradual, muddled socialisation of industries and services. Russia has not achieved more than we in spite of sacrifice of many human beings.*
> [Amélie at Aileen's Gilford General Practice to Dixon at Woosters, 13 December 1939]

Don Barron (American colleague from Dixon's time with HA Harris, now back in the USA) predicts, on June 12 (1940), two days after Italy's declaration of war against Britain, that 'if the allied powers can hold out for the initial latent period the resources of this country are theirs. Arms will of course go first but no one doubts but that is the beginning. Men would certainly follow'. But, can they hold out? Amélie

> *visualise[s] extreme poverty of course but that cannot prevent us giving them [the children] the rich inheritance of our lives – your knowledge – and my support in your doing so. The fear that is in my mind and which I express to you so that such self-indulgence may remove it, is that a Nazi conqueror divide us – take our children. The purpose of living would of course then go – still if one knows the worst possibility one at least can look at it. [Dixon shares the 'feeling that the survival of the children is the real thing ... I can always feel a sort of intuitive awareness of the fact that we, you and I,*

have <u>realized</u> ourselves in them more than in anything we ever could or will do'].

What plans are my friends making about their children. The Taylors [another Clare fellow] – Greaves – Woosters? I heard thro' Dorothea M[acdowell] [Belfast friend] that the Keirs [vice-chancellor, QUB] were asked to join in on some Cambridge boat [to America] for their children – do you know anything of this?
[Amélie, Newcastle Co Down to Dixon at Godwins, 20 June 1940]

'This' becomes the big question. Amélie finds herself understandably torn,

alternating between certainty that J and R cannot be brought up under fascism and absolute necessity for me to feel we are in reach of you. So I persue [sic] a business-like course in getting forms filled [for visa, money and transatlantic boat] backed by no decision or motive force. [If she does decide and if the authorities agree] ... the route which I can then best take – with Margaret [Godwin, to Canada] and to [Uncle] Charlie from there – or alone direct to Charlie is an open choice. The officials are so kind and helpful in characteristic N[orthern] Irish fashion. I have had a phone call booked to Marg[aret]t for 2 nights but so far have not got her ... The earliest sailing we c[oul]d get is approx July 15th ... [I was] at the American Consul Medical Exam to-day as the next is only July 18th.
[Amélie at Lennoxvale to Dixon at Godwins, 25 June 1940]

More philosophical thoughts on life, love and death, are sprinkled amongst the operational.

It heartens me to think you don't fear death – it takes the agony away and leaves a detachment and calm and eases me ... in my thought of you – which I seem to have day and night, each day and night because I LOVE you most terribly and long so for you. I don't, I think, fear death for myself – it seems so easy ... I am glad you love me. I want no one but you and in you I find, to give me pride, the things that matter so much the most, to me. Vision, intelligence and a queer sort of inspiration that is rare and valuable in my world of values – and then you see I love the corner of your lower jaw and your John's hair and your fingers and your vile temper and your stamen (as in Botany you know).
[Amélie at Newcastle Co Down to Dixon, 27 June 1940]

Dixon, in our memories, was indeed prone to explosive outbreaks of verbal "vile temper" though, unusual for the time, neither parent ever "smacked" us. His irascibility was balanced by a loving warmth and by a vision, intelligence and "Dixonish" inspiration which we increasingly came to value. What did "Examiner" make of the botanical, one wonders? (The censorship of inter-island post and Northern Ireland's different rationing – plenty of meat there – reminds us that a 'Border in the Irish Sea' is not new).

Bureaucratic demands on immigrants haven't changed much though the role of husbands [see point 2] may have:

The only hitches now for the Visa are
1. The removal from My Passport in L'pool of the Clara [her middle name] wh[ich] does not correspond with my birth certif. ...
2. A letter from you saying you give us permission to go.
3. Photos.
4. Kids birth certifs wh. I wrote you for on Mon.
4.[actually 5] Exit permit from Liverpool for me (as a doctor this just conceivably might be withheld).

I enclose a bank form. Kenneth Rae in B[ank] of Ireland said this a.m. 1. They wd give me a note allowing me to take £45 out with me and the children – 2. If you wd fill the enclosed form and send it to him personally direct K.R.Esq B[ank] of I[reland] Donegall Pl[ace] he will make application for monthly maintenance grant for me in Canada. If I do it there is less chance of success. This can be done thro[ugh] B[ank] of England and he is sanguine of success if we do not ask for excessive amount (say £360 p.a. – i.e. £30 p. month approx.). Can Marg[are]t not do likewise? Or if we go to Charlie cd. you work out any way in which we cd. help them – i.e. by you and Harry keeping a joint account ... By two days solid pursuit of forms, officials, and friends I seem to have made some headway.

I struggle to be more decided – but as you rightly said last night [a rare telephone call] I must make my mind up and my indecisions are only tiresome ... Everyone here is going round with indecision written on their face – asking anxiously – 'what are you doing' ...The Barcrofts, Dorothea, Aileen – all are in 'dithers'.
[Amélie in Newcastle Co Down to Dixon at Godwins, 27 June 1940 and 7 July 1940]

A few days later dithering is at an end as

> *we both agree and decide ... that if I get the papers O.K. ... I will*
> *book passages on the safest available route for the children and*
> *myself and go to New York.*
> [11 July 1940]

But, decision made, it was revoked for reasons which are not apparent. Amélie and the boys remain in Ireland for the next year and settle into shared life with the Barcroft family (Biddy and Henry and three children). Henry's research is on the very relevant topic of blood transfusion.

Shared life is both positive and negative. For Amélie, 'Henry is the best educationist of J[ohn]'s present environment. Each Sat-Sun he takes them [Boyd and Barcroft eldest children; both Johns] somewhere. Museum, electricity station. Railway. He enjoys himself like that as much as they do'. But, Dixon doesn't 'see why Biddy should not go to Newcastle [to allow Amélie a quick child-free visit to him]. After all I have been living without my wife for two years and I find it difficult to believe that Henry can't manage for a month or two'. And Amélie complains that

> *Biddy does thoughtless things. She brought Mickie down with a cold*
> *and throat week-end last and now I have had R[obert] out of sorts*
> *and J[ohn] Barcroft in bed all day and J[ohn Boyd] complaining of*
> *his tummy; it's obviously an infection imported by Mickie. She doesn't*
> *"see" germs and I do – one great difference between us.*
> [15 June 1941]

The families' friendship survived the war and continues to thrive into the childrens' old age.

Amélie's sisters wove in and out of her life of exile. Joan's marriage (awkward as Tom had been married before) was 'at our Reg[istry]off[ice] … At the Midland, they had some 12 people for champagne. Daddy went – at Joan's request. It wasn't easy for him and I admired the tough old fellow. He told Joan he wanted the bill from the party and behaved with great courtesy'. Peggy, unmarried daughter-at-home was a thoughtful support. She 'sent me Sarah

[Lennoxvale maid] down to-day. It's heaven. I must make hay and go to sleep'. The eldest also did her best; 'a parcel of books and puzzle from Abbot's toy shop in London sent by Helen came and gave much pleasure. Nice of her – and not like her!'.

John starts school. Clothes rationing arrives – no more chiffon. 'Nice to think your curley [sic] flannels outstripped the rationing ... after 14th [14 June 1941] you'll need coupons ... Clothes rationing seems an unimportant item in the struggle for existence ... after all you can do an awful lot with hats, lace(black) and blackout cloth (all unrationed!)'.

Unlike in England, food is plentiful and air-raids seem improbable though there are occasional warnings: 'three sirens in the last 3 days. The children have not wakened. They tried their gas masks on yesterday and would not take them off'. As for herself, mother and medical graduate,

> *I have washed my hair, poked the fire and am listening to the news + an alert. But kids did not waken. So, I must now compose myself and try to form the habit of not caring. Funny the physical signs of fear and so obvious that cerebral action is primary seat of bowel stimulation.*
> [9 January 1941]

Cerebral action or not, no major raids were expected. That expectation was in error. There is a raid at the end of March but for Dixon

> *it just is very bad luck that the first raid should have come after we had decided that things were safer there in the meantime. I don't see that we can blame ourselves and, in any case, statistically, and that is the only way one can assess these things I think that Belfast must be safer than any equivalently sized town in Britain.*
> [Dixon to Amélie, 9 April 1941]

It wasn't. A week later, during a single Easter night of 1941, Belfast suffered one of the worst raids of the war – 55,000 buildings damaged or destroyed and more deaths than in any other city except London. Amélie has been teaching first aid and shares a poignant vignette.

Y'day was to only a small class and quite elementary standard – but tinged with reality. A child of 16 came to me at the end – saying he had been with a friend for ¼ hour while he died in a raid and he had made up his mind then to learn First Aid. He was quite practical about it and undemonstrative.
[Amélie to Dixon, 17 May 1941]

One of the casualties seems to have been Dixon's accountant which does not unduly bother Dixon but does emphasise his, and governmental, concern with money and tax – war or no war.

Too bad about Brandon. I was just about to send my income tax business to him. Does his office still carry on? I would be glad to know by return if it does ... I don't need anything for income tax at the moment – in fact it looks as if I am going to have a little surplus this quarter ... The extras from Oxford have been very useful ... I should be getting a little extra for Primary Fellowship teaching this term and that again will help.
[Dixon at Clare to Amélie in Newcastle Co Down, 24 April 1941]

iv Wartime Cambridge

The London blitz over and Margaret's Cambridge offer accepted, August 1941 found us as a family in England living in 30 Barton Road with the Godwins. Four months later, Pearl Harbour ambushed the United States which, as Barron had predicted, then joined the fight. For Churchill, German defeat at El Alamein was 'not the end. It is not even the beginning of the end. But it is, perhaps, the end of the beginning'.

For us too, that year was the end of our beginning and, as the older two of us can remember, family life with inclusion of a father became a reality. Dixon, despite the missing years, proved to be an enthusiastic "Daddy". He kept Robert's *Otto the Otterhound* in the bottom drawer of his desk at work. There was a more grown-up volume for John. At bedtime he read us *Babar* but knew to skip for Robert the page where the mother elephant is killed; too like war. John was made of sterner stuff. After the move to our own household in Grantchester Road, we started piano lessons. For £2 Amélie bought a second-hand piano; it still plays nicely. The boys began to learn.

Adult talk was of war and sometimes of our disposal should Amélie and Dixon be killed. Though Miss Cooke, of Miss Cooke's Nursery, had offered to have Robert, the grown-ups felt it would be better for the boys to stick together. Peggy had promised to look after us if we were orphaned.

On February 18 1943, a few days after the German defeat at Stalingrad, Amélie was admitted to the Brunswick Nursing Home, a white building across the grass of Midsummer Common. It was her third delivery there. *Lying-in* for two or three weeks was the routine. 'Darlingest darlingest wife, just a note to say how much I love you – and your babies, I am hoping you will have not too bad a time [or worse? Death during childbirth was then rare but not very rare; some 50 times greater than today] Always and always, Your husband. Dixon'.

Next day, baby born, her father sent

> *hearty congratulations and all best wishes that the new arrival may thrive and his mother soon – but not too soon – be on her feet again. I know you would have liked it to be a girl this time but these things cannot as yet be arranged in advance.*
> [JMcC to Amélie 19 February]

Absent ante-natal prediction, such gender-talk, as from him or from a childhood friend, is prominent.

> *It's funny how you run to boys being one of a family of girls. Must be Dixon's fault! Or perhaps your father's strong wish for a boy got somehow tied up with your physiology. But seriously, I know you were feeling philosophic about it, and being disappointed for long about the baby's sex after it was born would seem like a sort of insult to the baby so I'm sure you aren't. He'll be sweet anyway and they are all so different. It must be fun showing him to John and Robert.*
> [Katie Brooke – childhood friend in Stamford – to Amélie, 20 February 43]

Showing Stephen may have been fun in her eyes. Robert had mixed feelings. In the bassinet on the floor, the baby's brown hair was prominent, the colour of chestnuts and very soft. John, now seven, and more relaxed at the new arrival, updates from the home front. 'I am awfully sorry that I have not written to you yet I hope you are getting on well and Stephen too. Kathleen [Irish maid] Nanny ... and Robie [*sic*] are well. I can not think of any more to say, love from John. This is a very Bad Letter'. Nora Wooster (no relation) was more direct, and very warming. 'I saw the adorable Steven[sic] Andrew in his basket. He looks a pet. No wonder you're so cheerful. Anyone would find him a tonic ... Love to you both. Yr affectionate sister Nora. PS first sister I've ever been really fond of!'.

By August, that fourth year of the war, Germany was in retreat in North Africa. "Second Front Now" was being scrawled on walls by admirers of Stalin. Bomber Command and the Americans were raiding Germany, we could hear their air fleets humming overhead at night; it really was a humming noise. The wireless told us that 'the *Scharnhorst* has been sunk'; good news, explained Dixon.

v   Anatomists as Family

After John got back from school – he was at proper school now – Angus Bellairs and his friend, Charles Shute, might take him and Robert for a walk. Angus wore a uniform. On leave before embarkation for the North Africa campaign (German retreat there not yet complete) he gave John a pet to look after; a grass snake which lived in its own box. Charles and Angus were to become members of Dixon's department. They were close to us all. For some ten years around the Second World War, Angus was for us an uncle-figure; for Amélie a surrogate son; and for Dixon a charismatic gadfly. As anatomists but also as almost-members of the family they deserve another digression.

> *I do like the boy, he is very enthusiastic, well read and has lots of naïveté in spite of his being a final year medical student who has swum with the cocktail communists for the past six years.*
> [Dixon on Angus, 16 May 1941]

That "boy" teased him intellectually in ways that remain both surprising and interesting. For example, in biology, what is the intellectual cost of becoming a specialist? Dixon enjoyed being his mentor – another anachronistic term – and Amélie agreed.

> *Yes, I'd have been pleased to have given enthusiasm to a student too. Great responsibility! Tommy [Walmsley] did a lot of inspiring the spirit of devotion to Anatomy – I remember the reverence you had for him and his pronouncements. I've always felt glad for you of friendships such as that of Bellairs. That eager enthusiasm for better worlds is charming and must make one responsible: or wish to debunk the enthusiasm! Really v. valuable stimulus. Can't you see us trying to live up to J[ohn]'s ideas of us – or debunking him fiercely in self defence?*
> [Amélie in Newcastle to Dixon at Godwins, 12 January 1941]

Angus and Dixon shared a love of reading as well as of obscure aspects of anatomy. He was a positive if unlikely entity in the lives of both our parents with the touch of an ever-youthful Puck to which they responded. From North Africa he wrote to them long letters describing the life of a young army doctor. The advance, as Germany retreated, then swept him up through Italy and, after victory in Europe to South-east Asia.

Angus had first appeared in Dixon's life in November 1935. A letter from HOM, to the 28-year-old Dixon just back from Baltimore, asks whether 'you [have] met a boy called Bellairs. 1st year Medical at Queens [Cambridge]. I saw him last summer and thought him intelligent'. Where HOM had met him is not apparent. The 18-year-old Bellairs had been at Stowe School [newly founded, distinctly "progressive"], where he had been taught by a young TH White, later to find fame with *The Sword in the Stone*, a favourite with us as a bedtime Dixon read. TH White, as the head of English at Stowe, took the intellectually more interesting boys under his wing. He was at that time obsessed by the sort of birds which led to his *The Goshawk*. Angus, the pupil, became obsessed by reptiles – 'my childhood pleasures when I used to catch snakes in the New Forest'. That teacher's private world seems to have included disturbing fantasies for which he sought help from a psychiatrist. 'Bennet is the name, initials E.A.' wrote White. 'He is a very great man – must be, because cured cases like mine are I believe most rare if not unique'. Wow! Eddie Bennet again. Amélie's cousin and Helen's long-term lover, analysed the man who led the Boyds' young protégée to snakes. Freud (or, for Eddie, Jung) would have approved.

Angus came from a prosperous ('in the wholesale trade') family but one whose fortunes had faltered following the '29 financial crash, as had his parents' marriage. The teenager, according to a late-life autobiographical piece, had read history, literature, and erotica; TH White surely the catalyst.

Scientifically, the development and underlying structure of different species – comparative morphology – was Angus's special love and snakes and their embryology the summit of his affection.

In North Africa his reading was not restricted to adolescent tastes. His letters remind that war leaves long interludes.

> I wonder would Dixon very kindly order 2 books to be sent for me from Heffers – one is Goodrich 'Vertebrate Embryology', the other Imms 'Textbook of Entomology' – if he can get them second hand so much the better. I am sending him a cheque for £4 to come express and suggest he sends them at an interval of one week so I am more likely to get one safely anyway.
> [Bellairs, RAMC HQ 4Div RE BNAF to Mrs Boyd, 26 Grantchester Rd, Cambridge, England Passed by censor 5795, 8 August 1943]

His general reading was as wide as Dixon's:

> Gibbon's *Decline and Fall* ('just the thing for the blackout'); Barbellion - *Journal of a disappointed man*; TH Huxley - *Collected works*; Thackeray – *Pendennis*; Walter Pater – *Marius, the Epicurean*, Ernest Hemingway - *Farewell to Arms* ('the most accurate warbook I've read'); Eden Phillpotts – *Saugusa*; Hakluyt's *Voyages;* Graham Greene -*The Ministry of Fear* [its Mrs Bellairs, the sinister fortune teller]; Robert Hitchens – *The Garden of Allah;* Compton MacKenzie – *Guy and Pauline,* Margaret Stern – *The sun is my undoing;* and [Russia on our side] *War and Peace,* Francis Cornford's *Poems from the Russian,* and Lermontov's *A hero of our time.*
> [Various letters from Angus Bellairs in North Africa to Dixon or Amélie, May 1943-June 1944]

Dixon and Amélie were both written to with no sense of hierarchy and no distinction between the scientific and the personal. Could they help with the problem of a wartime girlfriend?

> *She is a medical student at UCH, and should be qualified in about 3 months ... If you would like to get in touch with her, I should be very pleased. I have told her all about you. I think it would be v good for her to meet you and Dixon besides having the effect of freshening [her] memory a bit. I have no illusions about time and space, and considering it will probably be a couple of years before I'll be in England again, I am quite prepared for her to marry someone else in the meantime, which would be a pity I think. But I know Dixon's reluctance to meet anyone he doesn't know especially females otherwise I'd suggest he dropped into UCH refectory one lunchtime when he was in town, inquired for her and possibly took her out.*
> [Angus Bellairs in N Africa to Amélie in Cambridge, 10 August 1943]

She eventually came to Sunday lunch.

> *My dear Dixon, I've just heard from Pat that she has been round to see you recently. I'm so glad ... I gather that she enjoyed herself very much, was very struck by John and Robert. I think I must have held you up as something of a 'strong, silent man' – at any rate she seemed a little surprised – and delightedly so – by your talkativeness! ... I shall be very interested to hear your impressions of her. I knew her fairly well for about two years, but not as a lover, only as a rather detached admirer. Other things had to wait, unfortunately, owing to a combination of circumstances, until shortly before I left England. Incidentally she was the reason why I was only able to put in four days on the reptiles instead of a week! [Stamen's again?]*
> [Angus Bellairs RAMC HQ 4th (Brit) Div to Dixon, 18 Talbot Rd Highgate London N.6, 28 June 1944]

Could Amélie help get a paper of Angus's published?

> *Are Dixon's compulsions against publishing it really due to his dislike of wasting paper or having his name on a morphological article during a historical crisis? I've never heard of anything so fantastic – in fact I can hardly credit it ... I feel very strongly that 'pure' science should be kept alive at the moment: we are going to have*

such a fight after the war to prevent the doctors and other pragmatic folk getting complete control of biology ... I would be most grateful if you could chase D[ixon] up to embed and if possible cut [prepare for microscopy] some of the very interesting material that I sent off urgently ... the trouble is, as you know, that when I'm away, he is so apt to get completely involved in teaching and embryology and he will let the reptiles slide. After all, if I can collect these beasts in the middle of a battle under (admittedly very sporadic) shell-fire, he can surely cut them in the lab ... I've just wangled a beautiful holiday and got myself away from Divisional HQ to a Royal Engineers Bridging Camp where sappers learn to build bridges. I'm looking after about 250 'souls' instead of a thousand. So, my sick parade takes about 15 mins, and I'm free the rest of the day to collect reptiles. I managed to find Tom Day [Joan's husband] who is in a General Hospital five miles away. I looked him up last week and thought him delightful. He is coming to dinner in the mess to-night.
[Lieut.Bellairs, RAMC HQ 4Div RE BNAF to Mrs Boyd, 26 Grantchester Rd, Cambridge, Passed by censor 5795, 8 August 1943]

Angus was very close to Charles Shute. Charles was a Cambridge medical student when they met, having switched from English and moral philosophy (the claim by an obituarist that war service as an ambulance driver led him into medicine seems unlikely). The two of them were for Amélie 'almost a pair of twins'. After the war they both took up anatomy appointments at The London, under Dixon, and moved with him when he became the Cambridge professor. Charles remained but Angus did not. His wife (Ruth, not Pat of the lunch) who was to become distinguished in developmental biology, did not, despite Dixon's support, obtain a post in Cambridge. She did at University College London and Angus followed her back, to nearby St Mary's. It may have helped that a friend of Dixon, Frank Goldby, had recently returned from a spell in Australia to become head of its anatomy department. As herpetologist as well as human anatomist, the London Zoo with its snakes was, happily for Angus, close to St Mary's Medical School, the snake being, probably, a greater love than the medical student.

As far as I'm concerned, my interest in Zoology is easily the most important thing in the world and it means such a struggle to get into

a position to indulge it. Since I've been out here and seen the chaos which human beings can make so readily, I have become even more of a nihilist. Intellectually one can claim agnosticism only, but emotionally and by an act of faith as irrational as any Christian's I feel that human activity is so pointless that it really doesn't matter what happens to the species. For all I care, we can die out peacefully and let some other kind of animal have a shot at world domination.
[8 August 1943]

For Angus, that "other kind" would surely have been such as the "Zamenis specimen", collected in the last June of the war as the bloody break-out from the Normandy beach head was in progress a thousand miles northward. It was

brought by a military policeman who had the gumption to catch it unharmed – it was in the act of shedding its skin, and its new coat was a beautiful glossy black and white. It was nearly 5ft long – at first very savage, but later calmed down a lot. It ate three lizards (one was unfortunately a beautiful variety of Melanic Well Lizard I was going to send ... but stupidly put it in with the snake for convenience on a journey, never dreaming it would eat it under such cramped conditions) so I feel there is a good chance of it making the journey alive.
[Angus Bellairs in North Africa to Dixon, 28 June 1944]

Zamenis is thought to be the snake wound round, medical symbol, the staff of Aesculapius!

After his return from war, Angus came with us to Dorset snake-hunting. He wore a leather gauntlet. A slight rustle, a rapid lunge into the hedgerow, and the catch – usually a viper – hung wriggling from his grasp of the tail. The snakes were collected in an open tea chest in the garden of Mr and Mrs Neat's Swanage guest house. Robert was bitten by one of the rarer non-venomous smooth snakes; a pleasing show-off anecdote at school (with careful inspection he can still fine the hint of a scar on the finger involved). To our delight, we were also allowed to join excursions into Purbeck caves to "harvest" bats, another interest of the grown-ups. Bats could be plucked off the roof into a string-topped collecting bag. The Swanage beach was covered in leftover tank obstructions and barred to visitors but ice cream, a new experience, could be purchased at the North Pole café. Charles went snake-hunting with Angus too but his own research focussed on the human brain rather than reptiles. With Peter Lewis he devised a novel (and important) method of staining nerve cells to identify their *chemical transmitters.*

There was, for Dixon and Amélie, little or no distinction between the personal and the professional. Charles married, but his young wife, Pat – conceivably Angus's ex– was dying of tuberculosis. He came to stay, and mourn. After her death he too holidayed with us; not in Swanage but Amroth where Angus's head of department, fellow anatomist Frank Goldby, had Pembrokeshire family. With the Goldby children, we swam, fished, played and walked, and, when it rained, drank dandelion and burdock over Monopoly. The professorial fathers marked exam-scripts but, occasionally, could be persuaded to participate in cricket on the beach – French or with stumps. They would also converge in the pub when joined by James Bartley, along from Swansea.

Charles eventually married again. Their son was named Stephen, in honour we were informed, of our Stephen whose company Charles had enjoyed and of whom he had been especially supportive. A final third wife (his mother had had six husbands) was an Egyptologist. Retiring from Cambridge to Atlanta, that pair published a mathematical decoding of the Rhind papyrus. Charles, and Angus, were only two of several Dixon-protegees who spread their wings into successful intellectual divergence. He valued those with unconstrained minds

vi   Back in London

In January 1944, the evacuated London Hospital Medical School returned to London and we too left Cambridge, and Nanny, and our friends, and Grantchester Road. John had, before then, graduated from Miss Cooke's Nursery to Byron House School. To get there he, with David Godwin, biked all the way to Huntingdon Road pedalling past Miss Cooke's Cranmer Road.

Byron House had a sister-branch in North London's Highgate Village. A move there seemed better for the children than a return to Theydon Bois. The two older boys could start at that school with continuity of ethos for John. Adjacent Highgate School would, after Byron House, be a good next step (it wasn't, John hated it. He changed to the Hall School in Hampstead). But, would the journey to work be practicable for Dixon? He investigated; and looked at houses.

> *Going (by Kings Cross and Euston ie 2 changes) 25 minutes, coming back (by Charing Cross with only one change) took 35 ... [This] suggests strongly that Highgate is possible from travelling point of view. Cost return 1/4 [one shilling and fourpence] ... Went to 17 Cholmeley Park and got key from 34. The road is quite possible but very suburban, and the house is roomy, three rooms downstairs and four big and 2 small bedrooms. The garden however is only about 20 feet Square ... Obviously has not been lived in for several years. I don't think we would like it and it would require £300 or so to make it habitable. For £1500 it would be a good gamble but £2700 is fantastic. So, I think we can wipe this house off our list. I then went to Holly Lodge Gardens which is just beyond what I would take to have been the original Highgate village. The village is quite charming with many Georgian houses. There I saw first of all No 7 which would suit us admirably, hard wood floors, well planned garden, small but good garage but the price is much too high. Before the war I suppose it would cost about £2,700. If we could get it now for £3000 I would recommend jumping for it.*
> [Dixon to Amélie from London Hospital Medical College, undated]

They settled instead for 18 Talbot Road, Highgate. Less fashionable, and cheaper, it promised to provide a stable family home after five years of uncertainty and a manageable journey to Dixon's renewed academic life at the medical college. German defeat was now a realistic expectation. With the Blitz over, Russia advancing in the East, and the D-Day landing to come that summer, it seemed a good moment to say a final goodbye to Cambridge and to move on. It was not a good moment.

Within a few months "doodlebugs" [V1 and V2 German rockets] were landing on London (the very first exploded just round the corner from the medical college) and the boys, now three, were back in Cambridge; John with the Woosters, Stephen and Robert with Nanny. To Robert's sadness, Amélie remained with Dixon in Talbot Road only occasionally visiting Cambridge. To see her face appear at Nanny's kitchen window was a transcendental joy but felt infrequent. Duties assisting Dr Girling in a Highgate General Practice may have been a constraint.

First, the London itself was hit. The director of its Pathology Institute, distinguished pathologist Turnbull, described to Dixon how he,

> *with a sinking heart, ... went over [to] the Institute with a torch, because it held so much that was precious to me. But in the terrible shambles I found that the slide-cabinets with the sections collected since Jan 1, 1907, our records, all our paraffin blocks, the Staff microscopes, and even the trays of slides upon which Dr Woods and I were working were intact, and mine contained sections of three unfinished researches ... On no account write in answer to this. You work too hard ... It is a great pity that all the workers in the scientific departments do not lunch together as we did in the days of Bullock, Hill and Wright and, looking ahead it is one of the things that ought to be seen to after the war ... It is good and refreshing to suck the brains of workers in other subjects.*
> [HM Turnbull to Dixon, 23 August 1944]

A report to Dixon on the state of his department was similar: 'damage is limited to broken glass ... microscopes removed ... for safe keeping'. It was mentioned, in passing, that a nurse had been killed.

In August, JMcC had written to Amélie [14 August 1944] that 'Pegs [Peggy] feels strongly you should not stay in London but should go to Cambridge. So do I if your presence in London is not absolutely necessary, and if Dixon can manage – perhaps going too and paying occasional visits to London'. The advice was not followed and, on 23rd August, Amélie and Dixon were in 18 Talbot Road when a doodlebug landed on number 22. They were spending the night cramped into a metal-framed Morrison shelter which also served as their dining table. The Morrison protected them, like many others, from the impact of falling masonry and may have saved their lives. After the blast they could see blue sky where the roof had been. By coincidence, the Morrison had been devised by another Clare fellow, John Baker, using his "plastic theory of design". Decades later, after the death of his wife and of Dixon, that fellow became close friend of Amélie. We sometimes wondered if the closeness might evolve further; it didn't though his granddaughter recently described Amélie as a surrogate granny. Morrison shelters was installed in 500,000 homes. Provision was means-tested; Dixon paid for theirs.

Amélie told us that the blast had killed a bothersome marauding mouse which was 'good news'. A generation later, when PTSD had become a diagnosis, Robert asked her about flashbacks. 'No, nothing like that', but she did 'remember a neighbour rushing into the road from the house opposite bewailing the dead daughter in her arms'. That was not mentioned at the time.

Replacement of day-to-day items was a problem. According to Amélie's father,

> looters should be shot at sight. ... [You will need help with] immediate needs [such as] an alarm clock. We have only one that is used daily by the maids. A special permit is needed to buy one. You should be able to get a permit under the circumstances of your case. If, with a permit, you cannot find an alarm clock in your area, would your permit be available for purchase here? ... Bicycle pumps. Difficult to obtain but Peggy says she will send you hers ... Breadknife. We will try to get one.
> [JMcC to Amélie, 21 November 1944]

"War Damage" paid for building repairs but they took time. The next year was for the Boyds, as for so many, a time of renewed separation and of renewed support from others. Margaret Hill, a prominent local figure with a physiologist husband, offered use of the top floor of their house in Bishopswood Road on the grander side of Highgate. She had strong links with Cambridge. Margaret and Amélie became close and, as we shall see, remained so.

Although the rockets continued, John joined our parents at the Hills and restarted at Byron House. Robert and Stephen were left with Nanny and Robert returned to Miss Cooke's. Eventually, we were again reunited in a temporary London home round the corner from Bishopswood. It was from there, 11 Stormont Road, that, in victory, we crossed into Kenwood – burning rubber tyres, an enormous bonfire, and jubilant grown-ups – on Tuesday May 8 1945; VE day, the death of Hitler and the dawn of a new era. The next week another new dawn. Amélie left early from Robert's May 14th birthday outing to be delivered, at UCH, of her fourth and final Boyd Boy, Charles Adam Richard. We don't know why the first two names were never used. He has always been Richard. A calculation, 70 years later, suggests he was conceived during the week of nights in which the Morrison shelter proved its worth.

Monthly-nurse, Nurse King, in flowing uniform wheeled Richard out; "baby carriage", not pushchair. She found Robert's behaviour difficult. He, she complained, 'disgraced her uniform'. Employed for a few weeks, Nurse King was, despite her complaints, still writing to Amélie four years later – 'Love to you all, and thanking you' – a correspondence typical of Amélie as employer. At Stormont Road, Robert was kept in bed for measles (or was it mumps?). Dixon brought him paper to draw on which had swirly coloured pictures on the other side, 'what a baby looks like inside his mummy'. These were the proofs of *Human Embryology,* first edition, 1945; mustn't waste paper.

For six years after victory, we remained in Highgate. Amélie, mostly as mother; Dixon, as increasingly successful professor; we as children.

It was a time of austerity, for us again normality. After going to buy the sweet ration – two ounces its nadir, aniseed balls last well if you suck them – let's play on "the rubble" (the bombsite adjacent to 18 Talbot Road where we spent free time with friends, Nick and Mikey Deakin) till Dick Barton (Home Service 6:45, a not-to-be-missed quarter of an hour before bedtime). Rationing only came to an end after our final and permanent return to Cambridge. A return long-hoped-for as a possibility but also seen as a challenge to both parents.

7. Boyd boys 1 – on Hampstead Heath during family's final London period. From left: Stephen, Robert, Richard, John; about 1951.

3 A Cambridge Family

i   Appointment

In 1950 the expected retirement of HA Harris from the Cambridge anatomy chair must have been in Dixon and Amélie's thoughts and family chat. JMcC, in his penultimate letter to Amélie

> *wonder[s] what you and Dixon will decide about Cambridge. I am, of course quite unable to offer advice. You and Dixon will have to make up your own minds. I fully appreciate that the relative gain for D[ixon] in "otium cum dignitate", combined with other joint and family advantages, will have a very strong pull. On the other hand, the drop in income of £700, with a likely net budget deficit of £350 after savings as compared with London living expenses, is a very serious consideration. It has always been a strong point with me to live within my income.*
> [JMcC to Amélie at 51 Shepherd's Hill N6 (London house subsequent to Talbot Rd), 28 November 1950]

Amélie replies that 'Dixon [is indeed] inclined against it – school fees having gone up once again this term'. Educating four sons will be expensive.

By April the next year, Amélie and the two older ones are bicycling [not "cycling"] down the Loire – bridges still not rebuilt, few tourists, no rationing and strong expressed admiration for Churchill and the RAF – where, as victorious saviours, we are Anglo-special. Stephen and Richard have been deposited in Cambridge. 'Nice letter from Nannie [sic], boys well and sweet' writes Dixon.

Amélie, failing to connect with a letter at Tours, receives news redirected to Saumur:

> *Sat.21 April 1951*
> *My darling wife,*
> *The storm has broke! An invitation from the Cambridge Vice-Chancellor on behalf of the Electors – unanimous decision! A letter from Thirks offering Professorial Fellowship at Clare. Letter*

*from Le Gros pressing me strongly to accept the invitation. It would
have to happen when you are in France. £2350 + £50 family
allowance per child (i.e. loss of £400 per annum apart from extras).
I think, if you were here, we would agree to go, and I am making
decision in this way. Do say I am right even if you think it wrong.
Think of the advantages – life in Cambridge, the children there and
forget the awful six months move and settling. I do so love you and
would rather you were happy than that I were in Cambridge. But the
two, I hope, and believe, are not irreconcilable ... Imagine me in
Cambridge.*

In memory, Robert thinks John and he and Amélie, buoyed up by
the delight of those last four words, were all delighted too. It was
a culmination for Dixon, who said in another touching letter to
the Loire. 'I have literally shaken ever since getting the letter
[from the vice-chancellor] yesterday morning and can hardly
hold the pen – a sort of emotional hyper-adrenalism – I need my
wife so much in my crises ... I think we can be happy there – and
privately I will be more settled – there will be no place else to
go!' True, at least until US possibilities surface a decade later.

By August, Boyds – the whole family – are on holiday in
Amroth and Amélie is negotiating a place to live; for the ninth
time. She settles on St Chad's, 48 Grange Road, property of,
soon-to-be-lord, Professor Adrian, who, as new master of Trinity,
will be moving out. He writes to Amélie (handwritten, no typist
for domestic affairs of even the President of the Royal Society).

*What we want to do is to let the house (unfurnished) for nine years in
such a way that we could came back to it when I retire from Trinity
at 70. This may mean asking a rather lower rent, but the solicitor
thinks there is some way of preventing the tenant from insisting on
staying ...*

*The house is rather large. There is a garden which really
needs a gardener 3 days a week – we have had one full time but he
looks after the boiler and the hens etc. The top floor is mainly let off as
a self-contained flat. The people in the flat use the back staircase and
have a front door of their own ... On the first floor there are six
bedrooms and two bathrooms and on the ground floor three rooms
(dining room, study and drawing room) and kitchen etc. There is a gas
fired boiler for the hot water and a coke boiler for the central heating*

(there is also a supply of coke!). We have a wood fire in the drawing room and gas fires elsewhere when necessary. I don't know what rent we ought to charge – the solicitors will probably give me some idea of what it should be. The gardener will have the key and is there every morning and most afternoons, so come and look over it as soon as you like. I am not sure if the flat will be occupied or not. If they are there they could let you in. You will find my study in the normal state of mess. There are two garages, one for a small car and one for a large. I expect the flat might be glad of one if you don't want it.
[Prof E D Adrian at St Chad's to Amélie, forwarded to Amroth, 25 August 1951]

The Adrians moved out and the Boyds moved in, one medically qualified head of department succeeded smoothly by another 'Our cat seems to be reconciled to the [Master's] Lodge now, so I hope you won't have further visits'. We were to remain at St Chad's for half a decade until the family's final relocation, to 21 Newton Road.

On the day of the move to Cambridge, Margaret Hill, post-doodlebug provider of refuge, drove us children to St Chad's. John, boarding at Westminster School (many parental letters) was not to live there full time but he still had his own bedroom 'for the holidays'. Stephen and Richard shared a large one next to it while Robert was in the end room where, beside his bed, he kept a car seat pilfered from the ruins of a Talbot Rd garage. Covered in blue leather and very comfortable, it moved with him from house to house. Norah (Ansell, we will come to her), or Hedy Kerpen (Austrian child-carer of Belfast days) or, later, *au pairs* had a room round the corner on a corridor leading to bathrooms and a back staircase down to the kitchen and an outside door. That staircase led up to the top floor flat where Oliver Simpson, a junior research fellow at Trinity, lived. The door bore a notice: 'Do not ring this bell unless you want Simpson'. One particularly tiny friend of Richard's rang. Dr Simpson descended the three flights of stairs and somewhat tartly pointed to the notice. 'Why, if you wanted to play with Richard, did you press this bell?' 'Oh, I thought Simpson was the butler.'

It was indeed a substantial dwelling (now, more prosaic, it forms an offshoot of St Catharine's College).

ii   Visitors

It was almost rare not to have people staying. Many were relatives, close and distant, and friends from the Irish past. Others were anatomists from elsewhere in Britain or from North America. Many of the visitors became close to us children.

At St Chad's we also had "PGs" (paying guests) who occupied John's room in termtime and contributed to income. The PGs were language students from the "Bell School" a university-town magnet for children of the European rich. They played tennis with us and sometimes invited us to their homes. Jean-Marie's family owned grand hotels on the *Champs Elysées*. Robert's teenage night in a suite there was a pinnacle of unfunded reciprocity; he counted 13 towels. When adult John was posted as a diplomat to Germany, Reinholt, scion of a publishing family, might be a useful contact.

We addressed most adults by their "Christian names". Amélie's sisters were Helen, Joan or Peggy, not "Aunt". Anatomists were William or Marion not Professor Hamilton or Dr Hines. Such use, by children, of adult first names had, into the '60s, a 1930s feel with a whiff of coming from the political left. We didn't go the whole hog. "Mummy" and "Daddy" remained just that. But for the Woosters or the Moores, parents were "Peter" and "Nora" or "Mary" and "Terry". As we grew up and met our parents' professional friends in their public role our speech would sometimes need to swerve to avoid an awkward choice between "Professor", "Sir", or "William".

Visitors and their entertainment (and the use of hospitality to engineer social goals) loomed large for Amélie. Dixon [to John at Westminster School, 1 August 1955] is more ambivalent. 'Helen is here for Bank holiday – very Helen-ish but your mother seems to be able to cope! I am in the lab'. "Your mother" is a very common Dixon usage, usually a sign of semi-detachment from what she is up to. So is the escape to "the lab".

Amélie's social manoeuvring on behalf of others could be formidable. As Dixon reports, again to John,

the Moores arrived first thing that morning [December 27].
Dr. Heard, a fellow of Girton, had been invited by your mother to
come to see the Moores anent the young (I forget her name) Moore
and a Cambridge entrance. Dr H. treated this bit of skulduggery in
the right spirit. She came with three daughters and, as she is a widow,
her brother-in-law ... I had to restrain my deep, very deep, feelings.
And, after all, we were only 14 for lunch [after six other social
engagements in the three days preceding Christmas] ... so I retreated
into my innermost self murmuring ever so gently, donna strordinaria.
The Moores stayed until 8 pm which gave me the vicarious
satisfaction of observing your Aunt Helen's reactions when she
arrived at 6, expecting to find your mother at her sole disposal.
[29 December 1960]

We think Dixon did, in reality, enjoy visitors or, at least, half-did.
So did we, apart from the need to tidy our bedrooms to make room
for them while we moved in together. Acting as excursion
companions, as neighbours at meals, or as a "four" at bridge when
HOM came and being treated by most visitors as subjects worthy
of adult interest was an important feature of our childhoods.

Elizabeth Kinnaird – of the petticoats – was one loved
Belfast visitor.

Elizabeth and Richard

She was always affectionate to me when she had come over
to Cambridge; typically for a couple of days staying in the
spare room (Hedy's room) at Grange Road. Mummy was
pleased to see her and certainly wanted to encourage us to
be affectionate. The historical Belfast link to Mummy's
childhood where Elizabeth's father had been close to hers
was, I felt, an increasingly tenuous Irish connection for an
increasingly anglicised Boyd family. She had a good sense of
humour, deeply non-malign yet quite acute, as of the
stammering Catholic boy coming back from a job interview.
'There's - such - prejudice - at - the - BBC - against - Catholics,
so - of - course - I didn't - get - the position' (all said with
lengthy pauses between the stutters).

She was interested in art and the arts; not at all in
religion. Very much a spinster, 'one of the casualties of

the devastation of the cream of Ulster youth at the Somme,' said Mummy.

She, on each visit, brought one of her immaculately charming paintings. The water colour of autumn with rose hips and blackberries, that I think still graces the spare room at Robert's house, was certainly magical to ten-year-old me. And her hand-painted Christmas cards, from 1 Eglantine place - near Lennoxvale, across the Malone Road - were (like the Christmas Eve 3pm festival of nine lessons and carols at Kings) a brick in the spectacular, tottering edifice that Mummy, the agnostic, ever created around Christmas.

Elizabeth reminded me somewhat of Marion Hines, another visitor; intensity of interest in the young through depth of attachment to the mother. I never realised how far back the family link went. "JMcC" were initials of a brother so-christened in honour of our grandfather. An empty envelope shows he served in the Boer War. Did he die on the Somme too? He was never mentioned.

iii   Three Aunts

Amélie's sisters Helen (bossy) and Peggy (warm and loving) were frequent visitors, Joan (talented, clever, and absent) was not. They also deserve a digression.

Before Dixon, and before Elsa's collapse, the four girls' lives were not so different in tone from the ones we would have a generation later. Elsa, on holiday, enquires of Amélie if Joan is 'getting on alone with violin and Peggy [aged 12] with her lessons'. Joan may have 'failed in maths' at school but after leaving is enjoying Paris – including violin lessons – and is planning a holiday with 'Darlingest Am ... What date in June am I to reserve the hotel bedroom for? Where shall we go? St Jean? Italy??' She, artist-to-be, sketches family members in the margins of her letters. Helen plans holidays too, and shops. 'So glad you liked the dress: £4.14.6 – please. I had to pay for it on the spot. Won't it be nice wearing it all over Europe this summer?' [From

Ladies Writing Room, Peter Robinson Ltd. Oxford Street to Amélie, 3 June 1930]. In admonishing mode – a lifelong habit – she advises Amélie just before medical finals that revision must not be overdone: 'Darling darling darling ... I hope you are putting beautiful before clever which means bed at 12 – and not one minute later'.

Even before college, the three eldest have scattered. Helen, in her last year at St Felix School, Southwold, Suffolk, [10 July 1922] addresses her father in typically trenchant style. 'If you want me to stay near home I shall go to Dublin. If not, I shall go to London and get into one of the colleges there ... I would like to go away for a few months and go to college the following Oct. You see I do know my own mind at any rate – and am not floating around in an aimless way ... If nothing else turns up I can always turn into a ... private secretary for you – or anything else you may ask for. But Daddy dear – I am not going to fritter, Much love, Helen'.

Helen went off to Hamburg; Amélie to Lausanne and "finishing school"; Joan to Paris. Later, Peggy went to Lausanne too. Gap years, 1920s style, were followed by Bedford College for Helen and the Slade for Joan. Amélie, as we know, headed down the Malone Road to Queens. The older sisters gone, Peggy remained at Lennoxvale, often the butt of condescending letters:

> *She ought to get a holiday away from home ... Peggy wrote – was it at your dictation? ... I had a comic letter from P. [aged 17] today. I don't know whether to send it on to D[addy] or not. She went into the garden to look at the stars with a young man in the Air Force !!!!!*
> [Helen to Amélie 1930 and 1931]

There is an element of Cinderella in Helen's expectation that 'if Peggy is here the least she can do is housekeeping' or in Joan's comment, on a JMcC indisposition, 'don't let Peggy give him lobster and partridge together again'.

The lack of mention, in letters between these four girls, of their mother's illness and absences surprises us. Peggy was only 15 when the blow fell, Helen 24. Day-to-day gossip, thoughts for their futures, or reports on summer relaxation are the focus, not their mother. There is one rather searing exception. Helen has been dining in London with her father and Uncle Charlie.

> *Oh Amélie how hateful life can be! Well Charlie is away and Geneva [Pranguins] is in the offing and E[ddie] B[ennett] is being sent to look at it – and D[addy] is <u>terribly</u> wounded by Charlie's saying he had never understood M[other] and had been selfish – and that E.B. was no good. Of course Charlie was absurd ... the grains of truth were heated up till they were cannon balls! And it never does any good being brutal to people, oh lord!! I have just written Charlie to try and get him to be soothing to D[addy] on paper. Poor old D. You know Am. with all my sourness about the way he's gone on over this business my heart just bleeds for him (what an auntie ... phrase) but it's true. You know you can't suddenly realize what you are really like at that age without being shattered! Poor D. just sits about trying to count up all the nice things he ever did for M. [to] get back his superiority!*
> [Helen, no address, to Amélie at Lennoxvale, 18 May 1930]

They didn't write about Elsa and, a generation later, we didn't talk about her either. For those daughters, she must have been more than a memory, but not for us. Our recollections of Helen, Joan and Peggy do not include their mother.

Helen was tall, always rather smart and a bit intimidating. Robert had long agreed to her request, if demented, to 'polish me off'. But he realised, when she became confused, that to do so, both as doctor and as likely recipient in her will, would be doubly dangerous. He ignored the promise. Richard was more practical. She, desiring that life should end, also desired 'a drink'. Whisky was bought. He heard her reminisce, glass in hand, of 'the wonderful swing at Lennoxvale. It was in the large chestnut tree, and it swung high up over the bank, high above all those spring daffodils'. Post-drink, she fell, broke her hip, and died within the day – well done, Richard!

She was extremely sharp, rapier-like sharp in repartee, and unambiguously knew how life (at least the life of those of her social set) "worked". Her Irish birth and Ulster childhood were, in adult life, quite invisible. She was cosmopolitan and international in outlook and she appeared to despise the Linenopolis of her father and grandfather though she did admire JMcC's energy in furthering working-class education and health. (He had founded, we were told, the Belfast Council for Social Service). Claiming to be totally uninterested in her Jewish origins or in Jewish issues, she dropped an 'e' from her name solely, she averred, to make it easier to spell Lowenthal. (A cousin went a step further, changing hers to Lowell). Loewenthal and Iklé roots didn't stop her making the comment to Robert, when reflecting on the contrast between the poor in Hamburg and the lives of her Jewish banker relatives – long-since driven out and bereaved of a son-in-law by his regime – that 'Hitler had a point'. Her funeral was Church of England.

After school and Hamburg, she settled in London. Her degree at its Bedford College was on the language and literature of France and Germany. Then, at the Courtauld Institute of Art she both studied as a postgraduate and stayed on to teach the history of art. She also lectured for the WEA, the *Workers Educational Association* which provided 'Adult education to the Working Classes' before joining the Victoria and Albert Museum to initiate what would now be called an outreach programme.

John, who was probably closest to Helen, reminisced in a recording that 'Helen was the eldest and got out from Belfast as soon as she could. She was sent initially to St Felix's in Felixstowe [actually Southwold] or thereabouts. She clearly had an itch to get away. She came to London, but was just a very bright generalist, and so she was picked up in the war effort. All the girls, particularly Helen no doubt, had good German. Helen was very much involved with, not only things like propaganda, but in post-war preparations for Germany. And she knew everyone in that sort of world very well'. Richard wonders if she was a war-time spy. John, who mixed professionally with the world of "D-Notices", does not say she was, but he was always very conscious that "official secrets" had to be kept secret. The Courtauld had been directed, when Helen first went there, by Tommy Boase who was 'involved in intelligence' and was head of the British Council in the Middle East. According to JMcC, Helen started a 'secret job' in May 1941 that appears to have involved Boase and for which she was paid the substantial sum of '£480 per annum'. 'Is she still in Egypt?' writes one of her aunts. That Sir Anthony Blunt, surveyor of the Queen's pictures and notorious double agent in the Soviet interest, was, as post-war Courtauld Director, Helen's boss adds a certain piquancy to the question of whether Helen was a war-time spy.

John again:

> Post-war she worked at the V and A, where she tended to quarrel with people. But she had a couple of brilliant ideas, one of which was to set up this thing called the Attingham Trust [started, according to its website, 'by Helen Lowenthal, founder of the Education Department at the Victoria and Albert Museum, and Sir George Trevelyan'] which, in particular, brought rich Americans over to learn about the English Country House thing. It was a great success, everyone's always said. And this body she set up called "NADFAS" – the National Association of Decorative and Fine Arts Societies ... which is why she got her OBE.

NADFAS was conceived over Helen's dining room table. Now titled "The Arts Society", it is 'a network of over 90,000 people worldwide'. *Attingham* introduced her to the world of (wealthy) American museum donors.

> USA is grand fun. I've never been made so much fuss of. Parties are incessant and I am almost but not quite tired of being the guest of honour.
> [Helen at 1001 Park Avenue New York to Robert, 11 March 1966]

She talked less of her support for museum curators from behind the Iron Curtain.

John continues, commenting realistically on Amélie and Helen as a duo, that she 'had her flaws, clearly, but she was very bright, and she took a shine to me. Not having any kids, and liking to spike Amélie's pleasure in any way she could, she took me under her wing ... she was very nice to me, and clearly tried to get me to take an interest in higher things'. "Higher things" for Helen meant art and culture, not God; and introductions to individuals well placed to help access higher things.

Despite rivalry the two sisters enjoyed each other's company. On a snowy New Year, Amélie 'got as far as [87] Elizabeth Street [Helen's house] where Helen and I are amiably dug in. I bought enormous boots – must have used up entire sheep – and in these I'll attend Wyndhams Theatre (Heartbreak H[ouse]) and the sales (clothes Steve) and Wallace collection'. Helen had us to stay too. On leaving, leftovers from the fridge were produced together with thrillers or a Hemingway. Theatres and art galleries were a feature; and proffered friends and contacts; and advice, endless advice. 'Use a little Vim to whiten your teeth, smoking makes them yellow' or 'don't pull at yourself' or 'get a haircut' or, more weighty, 'make up your mind whether you are going to marry her so we know whether to be nice'. Recipients of advice were not restricted to family. Pevsner's biographer, Susan Harries, reports an occasion when Helen publicly chided that pre-eminent art historian for the absence of colour in his slides. 'I've always got on very well without, Helen' – 'You would have got your knighthood much sooner if you'd had colour slides'. Quintessential Helen.

Her Elizabeth Street, "smart", townhouse was tastefully decked out – Piranesi prints here, first editions of Hogarth Press there – with parties, for the museum establishment, both gay and straight. Rather surprisingly, her sole publication is a slight paperback *Gateway to Cambridge* published as an *aide memoire* for tourists in 1950. (£39 on Amazon. Dixon's *Human Embryology* at £41, bests it by a small margin). Amélie said Helen felt badly at her failure to author a serious volume.

According to John, 'Helen had lots and lots of lovers'. The most significant for her was Eddie Bennet though the affair was well hidden, at least from Eddie's son, Glin, who describes, in an autobiography, 'lunching with Helen in her exquisite house in Elizabeth Street, off Eaton Square. I had a spoon of soup in my hand just about to touch my lips – the moment is that clear in my memory – when Helen replied to my observation with the words "your father and I were lovers for thirty years" … the effect was cataclysmic'. Helen was devastated, said Amélie, that when Glin's mother died, Eddie married, *en deuxième noce,* his secretary rather than her. This was allegedly on the advice of Gustav Jung, Eddie's friend and mentor.

Aged 26, Helen had tried to reassure her father that 'although you see me far more in the light of modern fiction than is warrantable, I am really a thorough Puritan at heart ... don't you worry – I'll be alright as to health, morals, and happiness'. That prediction seems to have been close to being honoured over the decades before the final glass of whisky. She generally lived by one of the better bits of advice she gave Robert just before his marriage. 'Just remember what <u>fun</u> life is. No matter how many knocks one gets the plus is more than the minuses, it's – to coin a phrase – thrilling to be even the most average of creatures. Don't worry about anything – no greater waste and more tiresome activity than fussing about success, money, what people think and whom you have got to please'.

While Helen was often a doer by proxy, setting others to the task, Peggy was a doer by personal action. She was the daughter who fulfilled the role of housekeeper for lone father. She was also the young woman who faced down Belfast police or Queen's vice-chancellor in support of refugee academics in war-time Belfast.

Looking to make a contribution to reconstruction after VE Day, she, almost by chance, took a job with the American Jewish Joint Distribution Committee, also known, and known to us, as "Joint". It was to work at Belsen, Nazi concentration camp liberated by the British 11th armoured division in April 1945. Six months later she was there, working to enable its Jewish "DPs" [displaced persons] to reconnect with family if any had survived. The task was certainly harrowing, and it moulded her. She remained in touch with those she had helped for the rest of her life. She also got to know Egon Fink. He had had to leave Vienna following the antifascist Karl-Marx-Hof riots of 1934. Ski instructor in France (their introducer had been a mutual skiing friend) and French Foreign Legionnaire, he was now running a Belsen school for camp children. On marriage to stateless Egon, Peggy automatically lost her nationality and had to renaturalise as British.

Egon and Peggy moved to Morocco and both worked for Joint there. Peggy, pregnant, came to stay with us for the birth. Later, in Vienna, Egon was responsible for helping Russian Jews leave for Israel. John, being less discreet than usual, 'wouldn't be a bit surprised if he had been on the books of the CIA [or] worked for Mossad'. In old age, Peggy mused to Richard that perhaps the whole endeavour of moving populations into Israel had been a mistake.

In Vienna, the Finks' *Amalgergasse 8* provided a welcoming home for Robert when he spent summers learning German or working in a research lab of the *Allgemeines Krankenhaus*. Stephen and Richard also visited. 'We expect Steve in Jan ... off to Saalbach to ski on 23rd, Rick and [his friend "Gra",] Graham de Freitas join us'. Richard, on a subsequent visit, comments to John in a style which suggests older sister condescension has been osmosed. 'The Finks Haus is very lovely, Peggy very kind (and silly), M[ichelle] very young ... Egon talked, Pegs told me sweetly for the third time his life-story, how she enjoyed Israel, how she wished Ma would admit to her Jewish blood in a more open fashion'. This comment on Amélie's un-Jewishness is a surprise.

Condescension or not, this was the Peggy who, John and Robert had been reassured, would look after us if a bomb killed our parents and that did reassure us. Later, when Dixon was ill, she was, for Amélie, 'a <u>wonderful</u> sib'. Loving kindness was Peggy's hallmark and she brought it to the defining issue of their time – Belsen and all it meant. Condescended to, or not, it is she, of all the sisters, who witnessed for herself something of what is now history, at least as defined by mandatory goals for Key Stages three and four of England's National Curriculum!

Joan, born between Amélie and Peggy, was the absent one. In early life, Amélie had been competitive with Helen (it went both ways), and caring of 'wee Peggy', but was closest to Joan. Joan was also the Loewenthal who had been most accepting of Dixon in the early days. Amélie admired her but, in adulthood, things got complicated. Unmarried Joan was invited to visit Bentley Road. She would have an opportunity to admire baby John and Cambridge might also offer eligible young men.

One such was Angus, young, quick-witted, widely read, of the left, fun. He and Joan might get on but pushing him should not be too obvious? Our parents liked a couple who were at a similar stage in life to them and who also had a new baby. He worked in Cambridge pathology as did Dixon in anatomy. Both were medical academics much taken by their particular areas of scientific research and with similar university appointments. Both were, said colleagues, likely to go places. That couple were invited to dilute Angus at an evening supper party. Alas, Joan fell for the wrong man. The wrong man, Tom Day, fell for her. Angus did marry but not Joan, and not for another 10 years.

Despite the complication, Joan's aunts were pleased with her match to Tom. His having been 'head boy at Haileyborough [sic] is a mark of great distinction [writes (Aunt) Em Loewenthal, living at Fisher's Hotel Pitlochry, Perth, to Amélie, 2 February 1941]; a fellow is only made "head" of a big public school if he has character and moral worth … To be captain of the football team shows leadership and grit and determination … I just wish he showed more deference & attention to Daddy [JMcC]. Probably as "head boy" he was not accustomed to show "deference" to anybody!'. JMcC was unhappy at Joan falling for a married man.

After the war, Tom and Joan and their young daughter, Sarah, moved into 86 Talbot Road, down the road from our, now repaired, number 18. Stephen and Sarah played together and eight-year-old Robert escorted Sarah to Byron House. For the adults it was not so easy. Tom, in North Africa, had had, unlike Dixon, what used to be called a "good war". Post-war, Dixon, but not Tom, had a "good career". The two husbands didn't get on. To us it was implied that Tom was jealous of Dixon's academic preferment. Tom's son-in-law, by contrast, was told Dixon was a difficult man. Diverging family myths and diverging sisters, Joan did not visit. The froideur between the fathers was permanent and, while they lived, any relation between the sisters was equally chilly though Day-daughter, Sarah, at Girton reading chemistry was a link. In the '60s she was 'much [at Newton Road] … charming – tall and humorous … she bridges the … Day gap' wrote Amélie. Gap was the right word.

Dixon dead, Tom dead, and Egon dead, Amélie, Joan and Peggy came together living in Cambridge. Childhood intimacy was re-established but, by then, we had all long left home. Gifted, artistic, Joan kept on painting. Outwardly friendly to us, it was too late. We felt neither Peggy's warm affection, or Helen's bossy-cloaked love. Like all the sisters, she was over 90 when she died. Richard was surprised that her Grantchester Church funeral – '*Stands still the clock at ten to three*' – was also Church of England. As John has it: 'Joan was the least Jewish … as a type, and the least addicted to family stuff – she rather broke away'.

iv A Female Scientist

Marion Hines was a frequent visitor. Husband Leonard had disappeared early, and completely. 'He didn't like her becoming a professor before he did', said Dixon. Amélie thought the marriage had broken up because Marion could not stand Leonard 'always being late'; an odd comment from a mother who often kept us waiting. The professorship was at Emory, a leading US medical school. Marion went on to become a distinguished figure in American science, the first woman to be elected president of the American Anatomical Association, now honoured by a named lecture.

When Dixon finished his year at the Carnegie, Marion wrote to Amélie that 'it has been one of the joys of this year to have you and Dixon in Baltimore and to find that we all three could be friends'. After the war, that friendship thrived. She was a "regular" over many summers and a copious correspondent. 'I like the picture of you and Dixon drinking Irish whiskey at night together. I put you in the study before the coal fire'. Half a generation older, she was very close to Amélie, and to Stephen, 'my beloved Stephen'. 'Dear Amélie, what a delightful picture you drew for me with only a few words, of Stephen opening the package which held the book. The letter from him was truly wonderful. It made me happy for days, a nice and beautiful thing in the background. [She ends with a P.S.] Thank you for your letter Dixon … It never occurred to me that I would not like your boys. How could I not?'. Marion was, through her long life, interested in, sensitive even, to the young. She also became fond of our dog who 'was lovely even to me a stranger'.

A 1955 thank-you letter confirms that Amélie gave copies of *Period Piece* to friends.

> *Dear Amélie, Only two more nights on the Queen Mary and I can step off the gang plank ... the voyage has been very rough ... I have read Period Piece legitimately now and from the very beginning ... I would like to write Gwen Raverat a letter when I get home, but do not trust the Cambridge Post Office to discover one of their most eminent citizens. Would you put her Cambridge address in your next letter. Think of having a daughter so alien and full of protests. Could she have been as lonely and awkward as she pictures herself? I enjoyed the second reading more than the first. Think of having to chase your father up and down a field trying to get his attention in order to tell him you were engaged to be married.*
> [Marion Hines on Queen Mary, Cunard Line (posted New York) to Amélie at St Chad's, 15 November 1955]

v   Another Belfast Anatomist

We would sit and hull the strawberries which, in post-war short-supply, William, Professor William Hamilton, had "fixed" to obtain from Mrs Hacker at her fruit farm next to Cambridge crematorium. The ashes, said the professors, improved Mrs Hacker's crop. William's father had been, the story had it, commodore of the Larne-Stranraer shipping line. That is why John and Robert were allowed, when "crossing" to Belfast, to stand on the bridge and take the wheel, very possibly that of the *Princess Victoria* which later foundered with the loss of 133 lives; a great blow to the city. With birth of their twins, the Hamiltons trumped our addition of Richard; five children to our four. The two medical wives did not have serious clinical careers but sons did. 'Nice for William to have three children in the profession'. Again outnumbered, only two 'Boyd Boys' followed Dixon and Amélie into medicine.

For the coronation of the queen, William took Robert, with his older children, to stand in the Mall. It was 1953. As the Austin A40 speeded into London that dawn, the real excitement was his insistence we peer out of the rear window to warn of any police car checking our speed. This was the William of our childhood. Later as medical student, Robert was summoned to the dean's office. Had he transgressed? No, a message had been received from the dean of another London medical school, Professor Hamilton. Would Boyd care to come for Sunday lunch at his home? Oh dear, the clash of public and private. Should he now address William as "Sir"?

Dixon (and Amélie) and William had been close ever since QUB; a friendship extending beyond the dissecting room. In friendship, there is again a hint of condescension from Amélie (then still Loewenthal). 'William nearly fell in backwards when bailing the dinghy – like a large puppy'. He was a year senior at medical school but that difference never seemed to matter.

William was a prolific textbook author. Dixon and he wrote *Human Embryology* – Hamilton, Boyd and Mossman – together. The third author, from across the Atlantic and separated by war, played little part. In research, the afterbirth – the placenta – which connects baby to mother in the womb, was their special interest and placental structure a research field they made their own.

William was both a support and a goad in getting Dixon to bring matters to publication.

> *It is now 30th of December! William Hamilton arrived to work with me and I had to desist from my gossiping. We are revising the embryology book for a new edition, a process which I find less amusing than he does. He has a manic streak which keeps my nose to Heffer's grindstone ... William Hamilton has gone to South America, glory be, so he is out of the way for a little while.*
> [Dixon to John, two letters, December 1960]

As Dixon's health declined, William became an increasingly frequent working visitor. 'This next year [1962] William H. and I hope to complete our story of the human placenta which should be fun as we have some very nice, indeed unique, material'. William, driving endlessly to Cambridge, was loyal and energetic in enabling Dixon as he faded physically but not intellectually.

It was our then interpretation that William's "Roman efficiency" delivered Dixon's "Grecian creativity" (a popular comparison between the USA and Britain). Though we tended to assume the mental leader was Dixon, their contributions were probably more symmetrical. Was the slight condescension towards William that we sensed – perhaps imagined – merely Cambridge superiority, or was it rather, as with the Townsleys, to do with Amélie's more privileged origins? William finalised *The Human Placenta* after Dixon's death and, in posthumous fealty, ensured that authorship was given in the order Boyd and Hamilton, not the reverse. Amélie took consolation from an array of that volume in Heffer's bookshop window.

William was keen on the politics of academic life. When Robert asked him, long after Dixon's death, what he had most enjoyed in their years of collaboration the unexpected reply was 'keeping HA[Harris] out of the Royal [Society]'. His engineering of Fellowship of the Royal College of Obstetricians and Gynaecologists for Dixon was in character.

4 Childhoods

i   Schools

The letters stimulate memory of how our childhood was in the '50s, the decade starting from an Atlee government, and Korea, and the end of rationing through Suez, Hungary, and Macmillan's *"never-had-it-so-good"* to end in the "Swinging Sixties"; not a term we would apply to us.

Dixon and Amélie were now an established Cambridge couple and we established Cambridge children. Cambridge has always been "town and gown". We were definitely gown with the town close to invisible. But there were some differences between us. For John and Robert, class divisions were assumed, as were the boundaries they brought. A war shared by all weakened that culture. Stephen and Richard were less conscious of class. They were sons of the established professorial family without John and Robert's earlier experience as children of young parents making a niche as newcomers. For the older two, Dixon was a fit, albeit somewhat abstracted, father. Later, ill-health reduced his appetite for parenting beyond the academic. Academic parenting never lapsed!

All four of us were, by that decade, old enough to lay down our own Cambridge memories. School was, of course, a daily experience. College fellows' sons were educated, after "prep school", either in Cambridge at the Leys as "day boarders" – coming home for the night after evening "homework" – or, like John, were away, except in the holidays, as boarders at other public schools (that at least was our then-perception; selection bias in whom we knew may have played a part). The left-wing Woosters made a different choice. Their son, Geoff, was sent to the "Boys County". Tribe and class boundaries were high, Geoff's and Robert's lives diverged and friendship stalled; still a source of mild residual guilt for neglect of a shared earlier past.

John, at Westminster, was half-absent. He was not, unlike Nick Deakin, friend from the doodlebug rubble, a scholar *in college,* but he had an exhibition, the next best thing. Robert, terrified by the idea of boarding, refused to be considered for Westminster and started at the Leys. Stephen, after a brief and uneasy trial of Leys prep school, St Faith's, moved to less regimented King's Choir School (conveniently accessible through the fence from St Chad's) and thence to Bedales. Richard started at Newnham Croft, the family's first, and last, foray (until university) into state-funded education. It was a brief foray. He soon progressed to the Choir School too. Despite the name, and like most children at that, then boys-only, school neither Stephen nor Richard were choristers.

Richard later followed Robert to the Leys. It was not the parents' first choice for him. He is astonished to discover from the letters that, 65 years ago, Marlborough College had been preferred. No Marlborough place was offered despite five professorial letters from Dixon including the suggestion of an appeal to its trustees. That failure is a contrast to the entitlement we shall come across when university entrance became an issue.

We had no sisters. They would have gone to the Perse, slightly down market, or even to the "Girls County". Gender, like class, led to assumptions. As, unusually, Bedales was "co-educational" teenage Stephen was not cloistered within a single sex. The rest of us had little to do with girls in the years between early schooling (Miss Cooke's, Byron House and Newnham Croft all had boys and girls) and the start of teenage dances. Having no sisters didn't help.

ii   Transport and the River

Parents as well as children used the Cambridge vehicle. Its title had overtones. Nanny arrived on her *cycle.* Daddy went to the lab on his *bicycle.* We *biked.* 'Have you put your bike in the garage?' was a routine. Despite nagging, bikes were rarely or ever secured but possession of a "combination lock" was a prized novelty.

Richard's bicycle

Second-hand bikes could be bought on the first Thursday in the month, in the afternoon, at the cattle market, near Hills Road. Mine cost £10 second-hand. A bell was necessary; the law, I was told. I even received a hooter as a Christmas present, an innovation which, unlike the hula hoop, didn't catch on. Other Christmas presents included a small metal nut that when attached to a spoke on the front wheel, engaged a cog on a little dial and so clocked up the distance you had gone, and a device that could be attached to the frame of the front wheel and into which you could clip a tennis racket by its handle, though you needed to detach the press first (the press for a tennis racket was to me of utility similar to shoe-trees - nil. They were both considered to be essential for the care of the racket, of the shoe, and perhaps, like circumcision, of middle-class values). The clip could also be used to transport a hockey stick, but I didn't play hockey very often - it was rather too brutal for me.

Lights on bicycles were necessary; that too was The Law. My best friend's father was an MP (whatever that was), and Gra' had got into *big trouble* when he admitted that, yes, he had biked back from Rick's after lighting-up time without lights. Lights were detachable and needed 12V chunky batteries that had not run flat or been *borrowed* by a brother. So, the wrapped box containing a dynamo, run off the mechanical action of the back wheel tyre rotating the little dynamo hub, was an excellent Christmas present. Yet sometimes old people (Daddy would have been in his 40s) would reminisce of carbide lights - acetylene burning, with rather satisfying chemistry as you dripped water onto the white powder. But I only once saw one in action as I went over Fen Causeway bridge.

And there were useful backways that could be used. Routes for biking across Coe Fen, for example, that allowed you to cross over the Cam using the green metal footbridge with its convenient wooden trough in which the front and back wheels sat while you pushed the bike up and over. But the trouble was entering or exiting the Fen through the narrow *cow gates* that stopped that species wandering; a challenge to try and bike through without the removal of fingers. And the large cow pats on the path could be rather squelchy.

Daddy's bike was black and big. The crossbar was high as his long right leg swung over it. He was slow, yet he wasn't wobbly. There were no hills for him to cross on the way to the lab, but I noted in the evening that a headwind could make him somewhat breathlessly cross. And how did he transport things - books, papers - to the lab? No basket, nor briefcase that I can remember. And where would the bike have been placed in the lab? Was there a professor's bike rack? Hayward's bicycle shop was on KP, Kings Parade, opposite one of those colleges with Biblical-sounding names (Corpus Christi, Jesus, Christ's, Emmanuel), where exquisite house martin nests were to be found, beautifully sculpted under the architrave of the great arching college entrance. The shop later became Ben Hayward's which was not the same thing at all. You couldn't any longer buy linseed oil for a cricket bat.

Mummy's bike too had a crossbar, but lower; and it did have a wicker basket on the handlebar for carrying great bags of vegetables from the market next to the Guildhall and perhaps bottles of dry sherry, Tio Pepe. Her elderly friend, Mrs Hutton, was delighted when a bottle was *stolen* from her bicycle basket. This too was a sherry bottle but, on this occasion, Mrs Hutton was on her way to the doctor and had been asked to bring a specimen. This was felt to be a morally satisfactory yet mildly risqué story.

From a young age we 'biked' – to school, to friends, to outings – without parent or grown-up supervision. The same was true of journeys further afield. For unaccompanied train rides between London and Cambridge, Amélie would place small children in the train guards' van; a coin to the guard to 'keep an eye' on us. John and Robert, almost before their teens, might voyage unaccompanied from London to Belfast including finding their way from the

Heysham train to overnight cabin on the boat. More striking in retrospect is young Robert being expected buy his own ticket in Basel to travel solo across Switzerland for an Iklé-arranged skiing opportunity. A scrawled New Year's Eve postcard home from the 13-year-old reports that to purchase that train ticket he 'had to get money changed. [It was] out of date and the connection to Zurich was missed ... have to change at Zur[ich] for St G[allen]'. Teenage Richard "hitchhiked" with a friend to Skye. For travel, youth independence was encouraged and expected.

Nearer home, being on and in the Cam which, downstream, mysteriously changed its name to Granta, was a prominent activity. We all swam. Usually with friends.

> *Every morning at 7 o'clock Lawrence [Reddaway] comes to a certain window and tags on a string; that certain string is attached to a certain toe of mine! Having dressed we swim and walk and walk under water with or without a snorkel. I may see you at half term. November 5. Goodbye till next time, Steve.*
> [Stephen in London staying with HOM to John, about to start at Clare, 22 September 1956]

One memorable holiday, Biddy borrowed *Tiglath* – motorboat which had belonged to Keith Lucas, physiology hero of the "last war" – and we chugged down to Ely returning stern first as the mothers were unable to get out of the reverse gear.

Robert and his Leys friend, Sean, had canoes; Sean, the more able, could stand and punt, Robert, more physically cautious – a recurrent theme; he hated rugger and later, ski- mountaineering in Switzerland, held his companion back – preferred to sit. During a 1950s winter, Professor Jackson, Sean's father, took him and Robert to skate up the frozen Cam. We got as far as Grantchester (the father recalled that, as a young man, he had skated the 15 miles downstream to Ely). Smoothly gliding past the Meadows was exhilarating, and glorious but got Robert into trouble at the Leys. He was denied a day's holiday for breaching the school edict that we should not risk our lives on the ice. Sean, being with his father, was not; tricky diplomacy for its headmaster, our parents' friend.

Richard's gang

Things later went downhill - captaining the under-eights football team; just selected to be in the under-tens; then, rather smugly, never being chosen for any sports team. Similarly, my time as a gang member came to an end well before the, for me, vacant wasteland of school at the time of puberty. But the earlier years of gangland had been viscerally exciting.

The gang, the Boydani (Keynes, Rook, Griffiths, Hopkinson, Bertram - Cambridge surnames), had in common a lack of being choristers, and a very middle-class eight-year-old whiteness. We were all extremely small yet of course we were much braver, though less brainy, than Basil's Postani with whose gang members we often swapped allegiance. Shorts, grey woollen socks (purple at the top), aertex shirts, a grey pullover (not then called a jumper), and, in summer, sandals. All a part of a freewheeling happy childhood. I somehow realised this as I squeezed through the hole in the fence connecting St Chad's to the Choir School.

Ecstasy came when some of us found, underground in a corner of the St Chad's' vegetable patch, a mysterious seam of black shiny solid. It was asphalt, of quite unknown provenance but superb to melt in a large saucepan over a smouldering bonfire. The three most trusted members of the gang had been making a wooden raft, out of leftover bits of wood from packing cases. Suddenly we were able to waterproof it with a black shiny surface. Having added a mast, a discarded roller blind was a perfect Viking sail. We now had our very own ship, which we named *Stealer*. It was much too heavy to carry to the Newnham Mill pool, half a mile away. Luckily Keynes's family had a beautiful toy from Switzerland, a substantial wooden farm cart. We balanced and tied our raft on its top, but *Stealer* was still too heavy to push. Implausibly, the problem was solved by our teacher, Archie Keal - enthusiastic, young, sympathetic - who had a very small, very second-hand, Morris open-top with which he simply towed the cart, us three in the rear seat holding on to the tow rope, up to and along Grange

Road, along Barton Road through Newnham and on to the
Fen Causeway. There we placed *Stealer*, with her newly
painted burgee, into the Mill Stream. She wasn't very
buoyant (maybe too much tar) and was certainly very
wobbly when you tried to stand, but we did gingerly sail
her down towards the Mill where we moored her safely,
returning home in the tiny Morris towing the empty cart
to Herschel Road. Hubris was followed by disaster. Next
day the raft had vanished. It was only three years later
that I bothered to look properly right underneath the Mill.
There she was. I never told the others.

For those who could swim, the place to get into the river was
Sheep's Green, well upstream of the weir beside Scudamore's.
Gwen's *Boat-Picnic* in *Period Piece* has the boys naked at
Sheep's Green with maidens protected by parasols. In our time,
rough changing sheds protected passing mariners, and we wore
"togs", "bathers", "bathing-trunks". For Dixon, who didn't swim
in the Cam but did bathe at the seaside, costume was, at least in
our earlier years a photo reminds us, a full "bathing-suit".

The "little river" where John, and later Stephen and
Richard, learned to swim was a few yards away (Robert had learnt
in London). Amélie loved swimming too but, in Cambridge,
preferred the Jesus Green outdoor pool, also unheated.

Early morning swim

The tall Jesus Green pool attendant rather took to
Mummy's enthusiastic style of living - early morning swims
with me - or so Mummy thought.

She to him, 'What are the things, Michael, that really
give you pleasure in life?'

He to her, 'Well, seeing an old lady enjoying swimming
up and down the pool on her back.' (she might have been 54).

King's Choir School swimming in "little river" would
only be permitted (instead of afternoon lessons) if the water
temperature reached 60 degrees F, which it rarely did.

> Responsible Mark Hill and I (we briefly had a time when I would join him at Chaucer Road to bike together to school across Coe Fen) were asked to check the thermometer in the water at big river. I found that by peeing this could readily be manipulated upwards; I didn't feel at all guilty in this use of applied physiology. I don't think that Mark ever realised what I did, and why I was keen to be there just before him.

An outbreak of polio caused Clare fellow, virologist Michael Stoker, to make a fuss about his son being made to swim in Cam water. Sadly, little river outings then ceased.

iii   Reading

Books were central to the lives of all six Boyds and remain so for those still alive. Both parents read aloud to us – *Robinson Crusoe, Copperfield*, even *Bible Stories* (Dixon), *Lorna Doone, Secret Garden*, Joshua Slocum (Amélie) – and we read voraciously ourselves, Henty, or *Biggles* (the Empire was, for Henty, and John and Robert, a given; noble to be fought for and, for Biggles, a good setting for adventure) or Ransome or Nesbit or more adult works, perhaps at Dixon's suggestion. 'Ricky ... at the moment is immersed in Gibbon's autobiography. He, at least, listens to my ruminations. If I mention a book to Stephen, he avoids it like the plague'.

Amélie's reading was of the ordinary cultured variety. Biography, novels or poetry accumulated on her side of the bed. For Dixon, taste more diverse, reading was a core aspect of life, a devotion only to be equalled by love of what could be seen down a microscope. Volumes were read and acquired all his life. Many were second-hand, from David's market stall under its canvas awning or, on non-market days, from the shop round the corner. On Saturdays – the lab was visited six days a week, sometimes seven; on Sundays he 'slept in' until lunch then often 'went in' during the afternoon – the ride home invariably required a detour to see what was new at David's, jacket pocket bulging with successful *finds*. Dixon was also a serious customer at Heffer's, the biggest of the three or four large Cambridge bookshops that then existed. An invoice, for nine works purchased during a single quarter of 1943, includes titles ranging from the *Anatomy of Fabricius* to paleopathology. Two works were, as we have seen, for Angus on the North African campaign who later, in a Lancet obituary, confirmed that 'Boyd had an extraordinary flair for obtaining book bargains from the most unlikely sources. He must have built up one of the finest personal libraries on human and comparative anatomy in the country, and his range of literary interests extended far beyond the confines of biology'.

Dixon's favourite place to read was his private oasis, the smoke-filled book-lined study where three works might be read through by the end of a day. He would sit in a large armchair, the green one, with a heavy glass ashtray balanced to its left. Marginalia added, fly leaf signed, the book would be appropriately shelved. Subsequently he could, on the instant, find any desired volume and page to recall a point (unless some irritating cleaner or member of the family had moved it). Most volumes were hardback. Unless of French origin, "paperbacks" were not a feature in our earlier childhood.

Books might be in English, or in German, French, or Latin, or be mathematical. Many were biographies or literary criticism, others addressed early civilizations, the history of science or history more generally. And then there was the irresistible lure of Darwin and his followers. Many volumes addressed different aspects of Darwinism, some including the eugenic. Maybe Dixon closed his mind to the ethics involved but, if he didn't, he would certainly have been opposed to any censorship.

> *Biological facts are, very unfortunately, biological facts. ... [if we ignore this] it's a futile discussion.*
> [Dixon, Geology Department, Downing St. Cambridge to Amélie in Belfast. Examiner 2581, 23 June 1941]

Nabokov describes how, after his country's Revolution, he came across a volume containing his father's bookplate while browsing a Berlin bookstall. Dixon occasionally had a similar experience. The family suspicion was that these were books Helen wished to read and lifted from a Boyd bookshelf but, once read, fed into the second-hand market rather than risk discovery during return.

By 1957 paperbacks had arrived and another bookshop opened in Cambridge, Heffer's Penguin. It was in Trumpington Street convenient for the science departments round the corner in their Downing Street site and next door to Fitzbillies (that shop, to Robert's delight, had continued cake-baking through the war. When he went with Nanny 'to fetch the bread', he could rely on a treat, hot from the oven, supplied by friendly Miss Mason who served behind the counter. Stephen's earliest memory is from one such outing; 'my floppy hat with the sun shining down').

Heffer's Penguin was small and contained only Penguin paperbacks. On the day the Permissive Society arrived – the acquittal at the Old Bailey of *Lady Chatterley* or, more correctly, of her publisher, Penguin – Lawrence's then-notorious volume was on display. Stephen was somewhat dismissive.

> *There is now a copy of Lady C. outside Mummy's bedroom on the little table where things "to be done" repose. I agree with you that Lady C. is undoubtedly a book with a purpose, but I think it is badly written in lacking a sense of humour and a skill with words and sentences.*
> [Stephen to John at Yale, 17 April 1961]

Reportedly, the queue to purchase extended down Trumpington Street. Though not called to give evidence, Reuben Heffer, the company's chairman, had been on the list of potential witnesses for the defence. Heffers was publisher as well as bookseller. Its Mr Newman saw *Hamilton, Boyd and Mossman* "through the press" in "the Works" at the end of Station Road. There, he showed the older two how his machines turned paper into finished volumes with Daddy's name on. By 1948 that book was generating welcome royalty cheques.

> *3411 sold. Royalty £293-2-6 – Buy Mrs Boyd a nice spring hat with part of it please. Yours sincerely EW Heffer.*
> [Letter to Dixon accompanying annual stock report on *Human Embryology*, 7 April 1948]

"EW" was the company's then-chairman. His grandson was contemporary at the Leys and at teenage dances. Daddy strongly recommended Heffers when "old Joe Barcroft" (Sir Joseph, Henry's father, also a physiologist) enquired in 1945 about a publisher for his book, *Researches on Pre-Natal Life*. In the end, he chose the opposition, Blackwells of Oxford. Eighty years later Blackwells took over Heffers. At 52 Trumpington Street, the name of Fitzbillies lives on, also under new ownership.

iv   Other Mothers and Fathers

Social life in our Cambridge started young. In rationing times, a small bottle filled with milk was expected to be brought when we went to tea parties, and indoor shoes to change into. Only John and Robert lived that. As we grew, life with friends was mostly casual. If planned at all it was organised without apparent adult input. Mucking about in the garden – they all had gardens – going biking, or 'down to the river' were common after school or at weekends. So was just spending time with each other. In winter, one of the mothers might drive us to the fluid-filled fields of the sewage farm, an unlikely but popular place to skate during an icy snap.

We spent nights in friends' houses and they in ours. We joined them on holiday and sometimes they came camping with us. Nights in fields in tents were, thanks to her childhood days in the Girl Guides, an Amélie specialty.

For the older two, the social circle came exclusively from University Cambridge and, of university, the dons, not the technicians, secretaries or, except those at the top, administrators (Harold Taylor, fellow of Clare and the university's most senior administrator, was considered by all to be academic). A few major, non-university, individuals were, like the Heffers, also of the circle. Gwen Raverat's world of hierarchies survived; changed, but not out of recognition. But it is worth noting that Gwen's sense of hierarchy gradually diminished in force with the years and was less prominent for Stephen and Richard.

The houses of our friends were big and detached. Mothers were always present or seemed to be. As with Amélie, it was they who gave life to the home. Even if professional – as teachers, doctors, musicians, scientists, one a remedial gymnast, slightly below the salt – they didn't work, or if they did, only very part-time. Work by a mother outside the home involved a somewhat illicit absence from family duty. Their clothes were congruent with this perception; frocks, not slacks, at least until well after the war. For smart occasions, Amélie even had a pre-war hat with a whisp of a veil. Mothers usually did some of the cooking and childcare, and even a little housework but not most of it. Nannies, dailies and *au pairs*, who knew us and whom we knew did that. We also knew family pets, at least the dogs, and they knew us.

For us boys, the mothers were what, at least outside school, Cambridge was, and maybe we were right. The fathers might talk with us and play tennis or squash or take us skiing – not un-sporty Dixon – but it was the mothers who were central to our lives and to its ethos. Each was distinctive, they deserve to be recalled. Most had four children or, at least, three. Motherhood was an assumed good. Spinsters were to be pitied. Amélie's childless Belfast friend and distinguished social work researcher, Moya Woodside, who occasionally came into our lives, showed 'sour grapes' when she accused career-less Amélie of letting herself 'become a cabbage'. Sour grapes were to be understood in the light of her failure to have children or sustain a marriage.

Barbara Reddaway, mother to Peter, Stuart, Lawrence, and Jackie was the remedial gymnast. She did occasional work at the "open air school". They lived in Adam's Road just round the corner from St Chad's so we saw a lot of them. Robert and Peter discussed masturbation – minimise it – and religion. Robert tutored younger Reddaways in maths and taught the recorder, Lawrence came Amélie-camping in Scotland. He and Stephen conducted a vigorous correspondence.

Harold Taylor's wife, Joan, was mother to Micheal, John, Elizabeth and Judy. She knitted Robert ski gloves. The Taylors led Christmas skiing holidays for their four, taking along a large gaggle of other Cambridge teenagers. Across Europe by train and three weeks in Switzerland was an absolute highlight of the year. The gloves felt warm and affectionate, but Joan could rebuke:

'no wonder your faces are red this morning, the whole hotel could hear what you and John said last night about Sally'. Joan had been a teacher at, *crème de la crème*, Wycombe Abbey School, but now served at home, where John's room had an enviable collection of Biggles. Lovely, teenage-savvy Miss Stone, fellow teacher at Wycombe came skiing too. She had not married and was now a headmistress. Each of that pair had a path not taken; 'you can't do everything'. Taylor and Reddaway homes were slightly more formal than the Boyd's. 'Have you washed your hands' was a prominent pre-lunch query from Barbara and swearing was not heard in either house. Peter and John Taylor were cousins, we never quite knew how, and Reddaways joined the Taylor skiing party. There, fathers really did participate.

Helen de Freitas lived in Trumpington Rd, very close after our move from St Chad's to Newton Road. At Christmas, we all went round to sing at the "de Freitas Carols". She was interested in us, and loving and, like Gwen Raverat's (and Dixon's) mother, American. Hers was one of several families whose mothers brought a transatlantic refreshment to our lives. Helen was always a little formal in her hospitality. She did not have a profession but was, at least we thought she was, wealthy, which gave standing and had a certain glamour. Her fridge was the first one taller than us. Geoffrey was unusual in being an MP though almost an academic having been at Clare and Yale (a path John was to follow) and having college *dining rights*. Richard's closest friend was then their Graham, *"Gra"*. To Richard's sorrow, he went off to boarding school and to Amélie's later sorrow, the family ultimately left for Ghana. Helen's wifely support as 'an outstanding ambassadress for Great Britain in Accra, Nairobi and Strasbourg when her late husband ... served as High Commissioner to two newly-independent Commonwealth countries, and as President of the Council of Europe' [Obituary, *The Independent*, 17 December 1998] was in character. Before all that, Rick and Gra would take cake through to Geoffrey's study where pretty, young Betty Boothroyd took dictation. As a former Tiller chorus girl she was a change from nannies and au pairs.

Geoffrey dead and Helen old, Betty, by then MP herself and Speaker of the House of Commons, would come to cheer her up. Old in her turn, Betty later became the heroine of a Manchester Christmas musical! The de Freitas blue and gold chairs, seats from the Queen's coronation, and the framed Punch cartoon of Geoffrey as shadow air minister impressed us. Now, it is more impressive to learn that Geoffrey had some black ancestry, not something talked about in our Cambridge but perhaps relevant to his Commonwealth appointments. There were penalties in having a father in parliament. It was Geoffrey who insisted his son must have bicycle lamps which worked. Lawmakers' families 'must uphold the law'.

Biddy Barcroft, wartime co-mother and, like Amélie, both mother of four and a very part-time doctor was firm and straightforward and demonstrably brave. When Robert's penknife fell into the toilet ("lavatory" to us) she, without hesitation, plunged her hand in and lifted it out. This use of decisive action to handle an emergency was a life-long lesson. Although deeply Cambridge, Biddy and Henry did not live there. They were in Highgate to where, from Belfast, they had followed the Boyds. Both had had Cambridge childhoods. Henry's widowed mother, Lady B, relict of Sir Joseph, remained along from St Chad's in Grange Road at number 13. Biddy's father, a widower, fellow of Magdalen and, according to Dixon, 'a very distinguished mathematician of the tutorial type' continued at Howfield on the Huntingdon Road, a house nearly bought by us after his death. A Reddaway grandparent had lived next door to him; another link. As in Raverat, everyone knew everyone. There had certainly been occasional tensions during shared wartime parenting, but the Barcrofts and Boyds remained very close. Amélie spoke at Biddy's funeral half a century later and Robert, later still, at John Barcroft's, the other John from Newcastle.

Biddy, absolutely no fool, considered herself, perhaps correctly, to be the dunce of her Ramsey family. One brother, Michael, had been President of the Union and became Archbishop of Canterbury. Another, Frank, dead at 26, had translated Wittgenstein, and is increasingly viewed as a Cambridge economic and philosophical icon. The "Ramsey effect" is said to describe the impact of making an intellectual breakthrough only to realise that Frank had got there first. His widow became the Ramsey of Ramsey and Muspratt Photographic Studio, whose portraits – not those she took of us – live on in the National Portrait Gallery. Biddy was in and out of Cambridge seeing to father and widowed mother-in-law. The latter became, to use the parlance of the period, somewhat "gaga". On one memorable occasion she failed to answer an evening phone call so Biddy rang Amélie for help. Boyds, and a delegation of dinner guests, made their way along Grange Road, to number 13. There being no reply to Lady Barcroft's doorbell, a ladder was found and Amélie – doubly qualified as doctor and female – ascended. The response from the nightie-clad old woman on the upstairs landing was of the period. 'My dear Amélie how nice of you to call.'

Formidable, trenchant and somewhat scary, Nora Wooster enticingly left children to do as they pleased, only drawing the line when sex games with daughter Anna became too realistic. In a house full of exciting workshops – lathes, glassblowing, potter's wheel – Peter Wooster, again unusually, taught us to use them. Geoff set up a printing enterprise. Left-wing parental rules fined Geoff if he failed to buy to buy the communist *Daily Worker* when on holiday with us but were flexible enough to allow him to sing in a church choir. Their Soviet contacts enabled Anna to attend ballet school in Leningrad but failed to get them entry visas for a planned sabbatical in China. They then had to live in a caravan till their house could be freed of tenants; complexities of idealism. Despite Amélie and Dixon's earlier closeness to the Woosters, that family seemed, as the years passed, to be increasingly social outliers. Their network comprised the pro-Soviet, anti-American left rather than the Cambridge mainstream. Amélie and Dixon had given up on the USSR after it had 'betrayed us' in 1939, allowing Germany to focus on invading France. 'Not our war,' said Peter. That changed, of course, once Hitler invaded Russia. Dixon was trenchant in his criticism of the Soviets which he was also inclined to apply to Peter.

> *Molotov appears to have been just as unfortunate at appeasement as*
> *was Chamberlain. How one would laugh, if one could ... I see the*
> *C[ommuist] P[arty] have (sic) now stated that the present aggression*
> *really does justify fighting! Why, in heaven's name, can't these damn*
> *fools of human beings be logical and see that sauce for the goose is*
> *always sauce for the gander. Stupid men and women, with a religion*
> *and a theory.*
> [Dixon, Geology Dept. Downing St. Cambridge to Amélie, Newcastle
> Co Down, Examiner 2581, 23 June 1941]

The Wooster house, though at least as large as that of our other friends, lay away from the university at 339 Cherry Hinton Road. Their housekeeper, Janet, was, after the war, the only resident help we knew who was not an *au pair*. A passionately anti-American Scot and a single mother (Robert used to suppose a "Yank" must have been the father), she later eloped with another Wooster son, Tony, half her age; outlier indeed. Nora had, like other mothers, professional skills, physics in her case. Unlike them she used the skills in a masculine world as managing director of Crystal Structures Ltd., which she and Peter had started. (Though there had been some from the 1914-18 war, "spin-off" companies were then vanishingly rare, the term probably not yet coined. Today, it is claimed of Cambridge that 'around 61,000 people' are employed nearby by more than 5,000 knowledge-intensive enterprises with a combined annual revenue of over £15.5 billion'. The Woosters were early in the game).

Another mother, unusual in not being married to an academic, was Mrs Forbes; "Mrs" rather than "Evelyn". The Forbes family lived outside the town at Milton House, near the skaters' sewage farm. We don't know how they became part of the Boyd circle and they also, perhaps, remained slightly outside it and outside the university though Mr Forbes also had the right to dine at Clare. Having been district officer in colonial Uganda, they had come "home" when he was appointed curator of the museum at the Scott Polar Research Institute (her father had been geologist on one of Scott's earlier expeditions before the fatal trek to the Pole). Stephen remembers him explaining why Cambridge winters were so cold. They became increasingly close to and supportive of

Dixon and Amélie in the later years. They even asked a Boyd grandchild to be bridesmaid at a son's wedding.

Richard was devoted to another mother, Ann Keynes, whose son, Adrian, was the Keynes of *Stealer*, the homemade boat. A singer, she had married into Gwen's extended Darwin dynasty of which, and of her, more later.

v   Hobbies and Pets

John and Stephen played chamber music – for pleasure, not just as an extension of schoolwork.

> *Our 4 days music was a terrific success. We played three Brandenburgs and the flute suite, the Schubert octet, two Haydn symphonies ... admirable fun had by all, including the parents. Mummy's catering was, of course, fabulous, and we still haven't finished the bitter lemon. Other relics are some holes in the far end of the lawn, as used by a trio of two horns and an oboe. With the splendid violin teacher from school and John Gale, both of whom, with some other Bedalians, were staying with us, I sat up until midnight, playing p[ianoforte] trios – Haydn, Mendelssohn and Beethoven. Daddy sat in the corner and twitched! I was most impressed by my own standard, not having practiced for the last two years or so. Within two days I should get my A-level results.*
> [Stephen at Newton Road to John at Yale, 10 August 1961]

Robert and Richard, less musical, found other ways to pass the time. Robert, inspired by Geoff, had bought a printing press and printed cards or letterheads for friends and relatives. Richard, being the youngest, was just assumed to be happily occupied as, indeed, he was.

We all got involved with pets – dogs, cats, budgerigars, white rats, hamsters, guinea pigs, some doves and, once each, a snake (Angus's gift), a crocodile, a rook, a baby squirrel and a goldfish. The goldfish even lived on with a bite out of its dorsal fin, the result of an aborted assault by the then cat. The rats escaped to colonise hidden corners of the house, the hamsters a garden shed.

First in affection was Dinghy (Whiskers, the cat, was a close second for Stephen). She was John's. Dinghy is an odd name for a dog. It was chosen by John when he was given her at his January birthday in response to Robert receiving an envied "army-surplus" rubber-dinghy at Christmas. Regardless of formal ownership, and of species, Dinghy became a full member of the family endlessly referred to. She was a reassuring demonstration of family stability at the time and is now a powerful reminder of how we were.

> *Your mother called for me at the lab ... She, Stephen, Richard, Dinghy and I then went off into the wilds of Cambridgeshire.*
> [Dixon to John on school skiing holiday in Wengen, 6 January 1953]

> *Dinghy very frisky and accompanied Richard and self on a shining walk to Grantchester this a.m. at 8! We counted 11 fishermen with thermoses and far-away-looks sitting on the bank.*
> [Amélie to John in RAF (probably at Buchan Aberdeen), 4 December 1955]

She was fecund, very:

> *Main news here is Dinghy who delivered herself of six charmers on Sunday.*
> [Amélie to John at Westminster, May 1952]

> *Dinghy ... has been in heat and this staggers me! That sex is such a bugger – even to a poor old mongrel – why does nature want her to reproduce now? Anyhow we have been cosseting her. Hedy is seen with bits of chicken and I creep down in the night and even Pa is gentle and enquiring. She is sleepy and comfy and cocks an ear.*
> [Amélie to John at Westminster, 13 December 1952]

> *I think [she] must be pregnant – she certainly demands large quantities of food. A brown, debased Cocker the cause. Oh dear ...*
> [Amélie to John at Westminster, December 1953]

Following her death:

> *There is a gap in our ranks – not unexpected but as Rick says, if immortality lives in one's children dear Dinghy has that alright ... what a friend to have waiting out there among the Van Allen particles, along with limbs of previous unsuccessful Russian astronauts and Sir Joseph Barcroft's moustache.*
> [Amélie to John and ... his reply, March 1961]

Dinghy's successor, an unsuitably vigorous gift from the children to then sedentary Dixon was [Dixon to John 1961] 'a wriggling black puppy who soon grew into the rumbustious Leporello ... as black as they come; Neville Willmer [Clare fellow] has given his full name as Leporello [from Don Giovani] Grillo [Dixon's Nigerian research student] Boyd! And indeed there is more than a touch of the tarbrush in his composition. Moreover he has all the irrepressibility of a little pickaninny'. Note both the intellectualism and the casual racial stereotyping, absolutely of Cambridge and of the time. Amélie's similar description is of a dog 'black and very shiny with large Boyd feet and eye whites (Grillo type) and a tail like this [small sketch] too silly!'

Leppy soon grew large and wild. 'Has devoured, hens, rug, bedroom slippers – all one to him. Another day got shut in Rick's bedroom ... being bored ate a new paper basket, Rick's pig-skin stud box ... and the foetal skull ornamenting Rick's desk! And at my growl of displeasure all the starch leaves his ears and he is v.v. sad'. Both parents became deeply attached. While Amélie comments 'he is violent and so MALE', Dixon adds that 'bringing up four Boyd boys was a cinch compared to him'. He remained loved by Amélie, and Dixon, despite biting toddler-granddaughters on two separate occasions. Sons and daughters-in-law were not so forgiving. As with Joan, Leppy had come too late into the family for love to germinate.

We are surprised but, on reflection, not surprised, to note that almost every letter written to one of us from Amélie or Dixon – and many from us to them – include updates on dog, cat or budgerigar but especially dog. Recalling Dinghy provided a happy and shared emotional focus in writing where we all met on equal terms and with equal childishness. In the flesh, Stephen points out, Dinghy provided a soft shoulder to cry on when sharing a sadness with Amélie might have generated too much emotional ping pong. Lepporello's personality, and perhaps gender, did not fit him for such a role.

Stephen created an Animal Club with Richard; membership two. Eleven-year-old Stephen wrote on its behalf to invite Maxwell Knight, a popular radio naturalist, to an inaugural meeting.

Astonishingly, Knight duly turned up at St Chad's. We didn't know he was also a leading spy (his name plausibly inspired James Bond's "M"). Perhaps he chose to accept the invitation as acceptable relaxation during a visit to Cambridge to recruit undergraduates or, maybe more importantly, to talk with fellow spy, Victor Rothschild, who lived round the corner and whose loyalty was under debate; Knight feared, correctly, that the service had been penetrated by an unidentified colleague. Perhaps he just liked spending time with boys or with those who expressed an interest in natural history.

vi   Religion and Mental Health

Different ecclesiastical experience across our nine-year age-range is striking. Almost all Robert's friends of adolescence went, at least sometimes, to church. He thought it a Boyd eccentricity that our father self-identified as a non-believer and that we didn't go to church as a family. Richard is astonished by Robert's account and is certain that none of his friends or their families were observant. He early took against faith. We did all attend compulsory school services: Westminster Abbey for John and the Leys Chapel for Richard and Robert. At Bedales, there was "Jaw" from the head or a from an invited dignitary.

Upholding the family's atheist values could be awkward. Made to attend assembly at Byron House after initial exemption on (ir-)religious grounds, six-year-old Robert marched in swearing; use of swear words another family eccentricity. The more sober John was horrified. When, at the Leys, Robert failed to bend forward as others did for prayer, a fellow pupil assumed him to be "Mohammedan".

Richard's divinity class

Still vivid in my mind is David Isitt (young Kings theologian, early advocate of student counselling and founder of a residence for disturbed undergraduates). Asked, at the last minute, to take the divinity session for the Choir School class of eight-year-olds he rashly decided to take on faith. 'Faith is the substance of things hoped for, the evidence of things not seen'. Richard listened and reflected and realised there and then that faith was not made for him and vice versa. 'Never fall in love with your hypothesis' was, 40 years later, the title and scientific zeitgeist of a chapter in the *Annual Review of Biochemistry*. It resonated.

Dixon, thanks to a nonconformist childhood in America – no children's books had been allowed him on Sundays – had good command of Old Testament texts; 'stop behaving like a "bull of Bashan" ' could be an irritated shout if one of us became too noisy. His attitudes were not associated with any hint of Dawkins's *Selfish Gene* anti-religious stridency though Dixon did have a hatred of allegedly obscurantist Irish Catholicism. That scorn was shared by Amélie – and by Ulster Unionists.

Richard remains staunchly sceptical, Robert has become a regular churchgoer and John was somewhere in between; Stephen avoids discussion of this topic.

Then was the day, the heyday, of "Child Guidance", and of "Psychoanalysis", and the Boyds engaged. Robert and Stephen had difficulties, or were perceived by Amélie to have difficulties. It was critical that the opportunity, perhaps only a window of opportunity, for "therapy" should not be missed. The first nudge for action came, said Amélie, from Nora Wooster (creative revolutionaries – Freud and Lenin – sometimes shared admirers). She suggested Robert (aged ten) 'should be seen'. He has never been quite clear why; maybe childhood tantrums.

Miss Hutchinson's consulting room was upstairs in London's elegant Park Crescent. We were still living in Highgate and after a few accompanied visits he went by himself, on the 27 bus. Psychometric measurement was essential background. A bill of £2.2.0 enclosed with results provides the only evidence, beyond memory, of Robert's involvement.

Dear Mrs. Boyd, When I saw Dr Hutchinson to-day I suggested you might like a written report on Robbie's intelligence test.

October 23. 1945
Robert Boyd
- ○ *Chronological age 7 years 5 months*
- ○ *Mental age 10 years 11 months*
- ○ *Intelligence Quotient 147 on the Revised Stanford Binet Scale*

The results indicate superior intelligence (University Honours Standard)

Always interested in long-term follow-up, the subject is glad to confirm that the honours prediction was later fulfilled, though only at the 2.2 level.

Miss Hutchinson – Dr Hutchinson in the psychologist's report – engaged in enjoyable play-therapy interspersed with occasional attempts at interpretation. Robert concluded that Miss Hutchinson was not omnipotent when he hid successfully under her desk during consulting-room hide-and-seek; 'hiding in your mummy's tummy?' – the rising panic in her voice as she sought him, on the landing, down the stairs, into the street, was gratifying. Was it she who had the difficulty?

Robert's sessions came to an end but, years later, as we will see, he copied Amélie's approach, turn to therapy when a difficulty arises. In that case one he, rather than she, perceived.

By the time Stephen was 12 he was the concern and Amélie again swung into action. The nudge in this case, we are surprised to note, may have been from John, away on national service in the RAF. 'Thanks to you, I went to see Terry, whose letter will show you his sympathy and good sense'. Terence Moore worked at the Child Study Centre of London University's Institutes of Education and Child Health, perhaps as its director. He is the Terry of the Christmas visit already described. The Moore children had originally met Stephen and Richard at Byron House and the families became close. They camped together on holiday (though without non-camping Dixon). Stephen and Richard remember Terry warmly as musical, thoughtful and child-involved.

One concern seems to have been about homosexuality. Perhaps Amélie was worried by her knowledge of Uncle Charlie's lifestyle. Reassuringly, Terry felt [letter to Amélie 20 June 1955] that 'Steve ... is anything but effeminate or narcissistic, and has such wide and absorbing interests in outside things that I just cannot picture him turning into a 'queer' [Terry's judgement was correct!]. I very much doubt if any school can turn a boy or girl whose early emotional development has been normal, into a pervert, even though it may induce him to experiment for a time. The biological purposes are too strong!' But there had also been more general worries about anxiety, and about 'practical helplessness, forgetting, etc' so Terry makes a referral:

> *Stephen Coates, a friend of mine and a very well qualified Educational Psychologist, will see your Stephen at Gt Ormond Street on Saturday morning ... I shall be able to take him very conveniently. He will give him any necessary tests ... He will be glad to do this without any fee for a member of the medical profession ... He says he feels scared of the Professor ... I told him I thought it was probably mutual!*
> [Terry Moore at Child Health Study Centre to Amélie, 12 July 1955]

Amélie reports to John that

> *We were much surprised and rather perplexed to hear that [Stephen] had an IQ (for what it's worth) of 150 – and such deep inhibitions that this chap advised daily analyses for 2 to 3 years, living in London being, in his opinion, essential.*
> [Amélie to John, undated]

So, psychoanalysis would be 'essential', Terry suggests, to 'get him fit in good time before adolescence, and before he tackles GCE [General Certificate of Education]'. Note Amélie's 'much surprised' and 'for what it is worth'. She was always worried about Stephen and her comments often brush aside successes as in 'Steve is 1st in English and History (dud form but still...)'.

How to deliver psychoanalysis? London options, costs, and how a school might be found are debated. 'Would Westminster take him?' Or, 'failing a boarding school, could he not go to UCS [University College School] and live in a family from Monday to Friday?' According to Terry's wife, Mary,

> *Miss Freud [Sigmund's daughter, Anna] ... might put Steve on her clinic list; this would be definitely cheaper, but would not necessarily mean he*

would be with her personally but with an analyst under her supervision. The usual fee for non-clinic patients is £1.1.0 a session and they always want five sessions a week, Dr Woodhead only takes private patients, in New Cavendish Street. The only other place you could get free or not so expensive is the Institute of Psycho-Analysis in charge of Dr Winnicott [still a hero to many]. Terence would write to him for you if you wished. Also, Stephen Coates would write to Miss Freud for you. Fortunately, most of these analysts are now at a conference in Geneva. So, there is no immediate dash for you to write ... I can see that this will be a major revolution in thought and ways for many of you ...

PS Terry thinks Miss Freud's personal fee would be a lot more than a guinea.

[Mary Moore to Amélie, 22 July 1955]

Amélie sought the advice of other friends too. A letter from Biddy Barcroft, whose brother, Frank, had, as a 19-year-old, sampled psychoanalysis in Vienna, is typically straightforward. It illustrates the then-prominent role of analysis in intellectual-class thinking, and child-rearing. 'Am my dear ... feel glad you are taking a line' about Stephen. But what line? She has discussed the issue with her sister Margaret [Paul], an Oxford philosopher and another analytic enthusiast and enumerates.

1. *Are you sure that the person you took him to is in the first rank of psycho-analysts?*
2. *Margaret has heard a rumour that there is a psychoanalyst now in Cambridge [perhaps London wouldn't be necessary]*
3. *Some Freudian analysts believe very firmly that analysis is very little use during puberty and Stephen will be well into puberty when he has finished ... Mikey's Analyst [Mikey is child-neighbour and rubble friend Michael Deakin whose father, Sir William, is Warden of St Anthony's College Oxford, and whose stepfather, a senior BBC executive, was himself analysed] who we took [my] John to at 13+ years said she would not recommend analysis until he were 17 or 18*
4. *Your not being analysed would not matter at all at Stephen's age [wow!]*
5. *Analysis is by no means always successful – but as it seems to be the only remedy for a mental condition is worth trying*
6. *It never does harm*
7. *It certainly helped Helen [Margaret's daughter]*

But, she adds, Amélie should not 'feel guilty – 85% of children's complexes are born with them I'm sure and it's because you are such a good Mum this situation has arisen of your not being satisfied when others would have sat back and felt snugly contented with having produced such a charming boy as Stephen'.

Strikingly, she mentions [13 August 1955] that analysis would be 'against Dixon's wishes' before concluding that 'Bedales might be just the place … If you are very worried about a child that is another reason for sending them away from home, as you don't worry nearly so much if they are away and your anxiety is very bad for them'.

Charles Shute disagrees with Biddy's assurance on safety. He is

> *V[ery] set against analytical treatment of which he maintains to have much experience – and thinks it dangerous and only to be used as "last resort".*
> [Amélie to John]

Softening what may have been said to Biddy, Amélie adds that 'D[ad] has, for him, been wonderful' and, in disagreement with her comment on Stephen's dud class, reports that she has 'interviewed old Butters [Choir School head] on Steve's behalf and arranged with [him] to coach S. for Marlboro' scholarship'.

In the event, psychotherapy and psychoanalysis were, for now, abandoned. Stephen left the Choir School in due course, not for the Leys, nor for Marlborough, but for Biddy's Bedales. Terry's career thrived and he became a professor himself though not in Cambridge. John, reflecting in old age to daughter, Jessica, has a straightforward view.

> *Amélie … had an absolute obsession with Freud – she always wanted us all to get analysed. And the only one who went for it was Robbie [later]. And actually, Robbie is so transparent you can analyse him from outside in minutes, but he spent a lot of money, I mean he wasted a lot of heritage, I think, on deep analysis. With Uncle Steve I don't think the issue ever arose, since he always lived in a world of his own anyway. Uncle Rick thought it was a load of cobblers [John wasn't quite correct; Stephen, too, later had analysis].*

123

vii Boys and Girls (and Sex)

Following middle-class rather than Jewish culture, we four were all circumcised as babies. Richard's rather grander contemporary, King Charles III, was also done, but none of Amélie and Dixon's five grandsons have been thus ritually disfigured. Not exactly sex education but, at least, a comment on evolving attitudes in regard to the penis and its works.

Apart from anatomical dissections or genetic tutorials with Dixon and warnings not to make a girl pregnant – not *how* to – from Amélie, our evolving adolescent practices were topics they preferred not to talk about; surprising, retrospectively, in the light of their letters to each other – stamens forsooth! Facts but not feelings could be discussed when sex was in question, Dinghy a useful bridge. In school years, a gulf of silence on personal matters of sex was maintained by both parties, dancing lessons, parties, and favoured female invitees could be discussed but not below the waist.

In Robert's, but not Richard's, experience, open communication between male adolescents was almost universal. Frequency of masturbation was a major topic. While "flogging" (local jargon for the practice) was no longer a plausible cause of blindness, it was widely believed in his time at the Leys that its performance used 'as much energy as a five-mile run'. Abstinence might be prudent. Performance is the right word. Masturbatory techniques were – literally – handed down and then practised together. One feisty Leys' boy claimed to have had sex of this style with half of those on the school roll. Conversely, of a "sex scandal" at Oundle School, Leys head, Gerald Humphrey, shared with Amélie his relief that 'there is nothing like that at the Leys' – gaze avoidance indeed!

Richard's remembrance of sex at that very Leys seven years later are completely different; none. Had there been a clean-up or is there selection-bias in our different memories or was there just a cultural shift? Stephen has no prominent memory of unorthodox sex at Bedales though a rumour of one boy's practices sticks in the mind – 'Plum, the bastard'.

Robert's extra-curricular education

Adolescent experiences outside school were several and are recalled with absolutely no sense of victimisation; inconsistent with current concepts of "abuse". Discussion with some male contemporaries in later life suggests that such experiences were probably not untypical. None, with the possible exception of ten-year-old John being once assaulted, by young Wykehamist, Dorothea McDowell's Christopher, had any element of violence.

Learning started young. Godwin son David, 11, went with John and me, 11 and nine, on an excursion to a Harry Godwin friend, a school teacher who had left Cambridge to teach in Windsor, presumably at Eton. It was a marvellous day; tour of the town, an inspection of the young man's rooms including his bedsit and, accompanied by our, or at least my, delighted giggles, "tickling" on the bed not excluding testicles and penis (yes, those were the words used in our anatomical family). John was less enthusiastic. (Bad memories of Christopher may have influenced his view. Certainly, he later disapproved strongly of David's father, suggesting years later that Harry had been complicit). A delicious tea followed and a suggestion from me to repeat the tickles completed the visit. This learnt activity being so pleasurable I shared it - in his parent's garage - with Byron House friend, Peter Weiss. He was a good pupil.

Another experience was with ex-navy Commander Palmes (C.P.), a figure in Andermatt where the Taylors took us skiing. His seduction pitch, over skiing magazines, was a very Victorian injunction: 'never rub alone'. Victorian doctors had (or pretended to have) complete confidence that masturbation, solitariness implied, could lead to blindness, imbecility and worse. CP spiced his health advice with generous financial and other gifts and his fate nicely illustrates how a gulf of silence could be breached - follow the money. I had to keep a financial account of holiday expenditure for Mummy. Cross-questioning about an unexplained surplus disclosed a gift from C.P. and thence its cause. Little comment was made to me, but Harold

Taylor, tipped-off by Daddy, expelled visiting C.P. from the Taylor home. Taylor-son, John, had previously stayed with C.P., been given a wireless set by him and had had the chance to ride his motorbike. Robert was envious and remembers pressing John to write, post-expulsion, to C.P. suggesting we might go together to visit him. Prudently, C.P. did not reply.

Harold, in a letter a decade later, thanked me for saving John; an undeserved approbation. Even now, I feel mildly guilty on contemplating this two-faced behaviour towards, in modern parlance, our groomer. That I also felt so 60 years ago is implied in a comment to (brother) John in response to that letter. '[Harold] wanted to thank me for noble behaviour in exposing C.P. - John's pederast whom I so much enjoyed going to bed with years ago. I shall mark this down as one of the highlights of my acting career'.

Robert's experience was not unique. Donald Butters, Stephen and Richard's popular head at the Choir School, suddenly evaporated from his role probably, said another pupil to Richard, because of sexual goings-on. Nothing was made explicit. Two generations on that school now has a female head and girls in the Choir.

By the time of school-leaving, John was well into girlfriends but Robert was more excited by boys. When new students at Cambridge were invited to share any health concerns with the student medical service, he mentioned this worry. The news was, with his agreement, passed to the GP who promptly, and without his consent (another culture which has changed) told Amélie and Dixon. There was a sudden, awkward, and partial reaching across the gulf. 'I always thought your friendship with Pringsheim [an older school friend, who mentored me when we biked home together after school and whom, while admired, I did not fancy in the least] was a mistake' or the 'I shared a bed with Norman when I was a teen-ager, there was nothing like that'.

The issue was handled by Amélie and Dixon as it had been by me. Sexual desire was a medical matter. In fixing for me to see Eddie (the medical cousin again) and in accompanying me to Harley Street to do so, Dixon was entirely supportive. Psychoanalysis did, this time, ensue. Robert, reflecting after more than 50 years of (straight) marriage, wishes to share a deeply unfashionable conclusion. Childhood lessons such as the ones he received (and the ones he gave Peter Weiss) lead him to be sceptical of the idea that sexual preferences cannot be both taught and, once taught, be modified by different experiences.

How boys met girls, or rather how the genders (not then a category outside scientific discourse) met in formal array, is a prominent theme in *Period Piece*. Our experience was more diverse. In John's (and Robert's) adolescence there were still holiday dances with formal invitations. Dancing classes were attended as preparation – foxtrot, waltz and quickstep – 'you will enjoy the parties more if you can dance properly'. From, single-sex Westminster, such events were a big deal for John; Frankie, or Jill, or Catherine were all potential. The question of whether one of them would become a fixture remained prominent in mother-son correspondence into the next decade (letters found in the suitcase from a young JMcC to our great-grandmother have a similar flavour; family behaviours tend to repeat). For Stephen, at "co-ed" Bedales, day to day issues were more important than holiday dances. In the middle of A level exams, he is

> *having a terrible and one-sided affair with an extraordinary girl ... has me right at the end of a string – she is unnervingly attractive in her behaviour and actions, though not really beautiful in any classical way, with a sort of the defensively pugnacious way of treating everyone – me included! In fact, I've got it so bad in my belly – you know the way I mean – that I can't concentrate on my work. I reckon I've got more than my fair share of Adrenaline ... It's bad during the exams, and I have to get up in the night and go for midnight bathes etc to get it out of my system – you know, the classical adolescence, but I'm not at all sure I want to get out of it into the sheer defeatism of most adults – it's a difficult choice, especially as I've wasted my education up-to-date.*
>
> [Stephen at Bedales to John at Yale, 2 July 1961]

By Richard's adolescence, navigation of teenage social life was rather left to him. Amélie, more relaxed, reports [3 April 1961] to John 'a change in the wind at the Leys – a boy sacked for being <u>caught</u> in animal delight. Minnie Moore [head of maths] for being <u>caught</u> in flagrant delight [with a female teacher] – a prefect appointed for brain not brawn'.

viii Sons of the Professor

As children, Clare belonged to us and so did Dixon's department, "the lab". On London Saturday mornings, the older two accompanied Dixon to Mile End Road, doubtless to give Amélie a break. As he worked, he would occasionally invite us to look down his microscope. We read and wandered round. In the "museum" was the deformed and bottled head of Joseph Merrick, "the Elephant Man", who, until his death, had been given Victorian sanctuary at the hospital. Long after we had paused to shudder at that "specimen", Merrick became the unlikely hero of a 1980s film, then of a 2023 play. (When Robert tried to take his son, having seen the film, to inspect the head, it had been withdrawn from display; a further, somewhat squeamish, example of cultural change). In the lab was an X-ray machine which Robert played with to image the bones of his hand (did I wear Dixon's protective lead apron? I don't remember). One machine was Dr Bourne's Geiger counter, unusual technology in anatomy. He left from Dixon's department for the USA and became, at Emory, Marion's head of department. She, friend of Dixon, came to loathe Bourne.

Biological education, during those Saturday visits, was an odd mixture. Bits of bodies sat around in pots. 'How do babies grow?' 'Come over here and I'll dissect a foetus to show you'. 'Women live longer because their sex chromosomes are paired unlike those of men', a relief because death of Mummy was an ongoing war-time fear. Dixon seems to have assumed such topics had been absorbed. When a letter later arrived from the Leys encouraging parents to explain 'the facts of life', it was merely handed over with an embarrassed 'you know all about this don't you?'. Robert didn't, despite the Saturday mornings. He only understood the anatomical nature of "the facts" once he himself could, as a medical student, dissect an adult female, albeit a dead one.

To us, Dixon was owner of his department whether in London or in Cambridge. We, his sons, used the departmental darkroom to develop our holiday snaps and had departmental white rats as pets at home. Its chicken eggs, having had their embryos removed, were war-time scrambled egg. We had free run of the "animal house". Our friends would wander round the department with us and would sometimes be given Dixon's help. Under his supervision, Richard's Adrian could dissect a brain or David Godwin have a tutorial. Members of other departments often reciprocated.

We were expected to address technical staff appropriately, *Mr* Fozzard or *Mr* Thurley. For Dixon, they were "Fozzard" or "Thurley". With Fozzard, Daddy broke the class-barrier to the extent of enabling him to dine at Clare. He also told Richard that Thurley was so able he should have been the professor rather than him. Did he mean it? (After Dixon's time Thurley began to publish but we don't think he ever became professor!). Social class lingered on. So did gender. His secretary was *Miss* Salisbury to all the Boyds and, as Richard discovered, to others as far afield as Oxford.

An Oxford Professor

'Boyd, you are Boyd aren't you?' said the professor. This was November 1966, 8:15am on a Saturday morning, a suitably committed start to my own research career. I was doing an experiment in the brand-new biochemistry department, alone, on the very bottom rung. He, Krebs (Nobel 1953), was out of sight at the very top. This elderly professor who would retire at the end of the academic year (my current decade would now call him middle-aged) was doing a round of his department, snooping really, to see who was at work at that time on a Saturday; not many were. I confirmed that I was Boyd. 'Well, give my greetings to Miss Salisbury, please.' She was our father's (wonderful) secretary. When Krebs as a young refugee from Freiburg became, in 1934, a demonstrator in Cambridge biochemistry he must have got to know her - she had worked originally in that department. Krebs must have learned, somehow have found out, that Dixon Boyd was now her boss (and that I was his son). She never married.

At Clare, Dixon was member of the college family but not in charge. Thirks – "the master", Sir Henry Thirkill – seemed to have that position, the one that JMcC had at Lennoxvale; if not sole owner, certainly college grandfather. For us, child-members, it was straightforward. The fellows' garden was ours; the tennis courts, our sports facility; the fellows' library, where we were taken to be introduced to and handle incunabula, books printed before 1500, not quite ours, but also in the family. Other Clare children were – ex officio – our friends. They ate with us and we with them. They often stayed. We holidayed together, played together, grew up together and experimented together. Experimentation was not restricted to the sexual. Robert remembers anaesthetising himself, and perhaps some of them, with chloroform having first removed impurities with (was it?) silver nitrate as recommended in a 1920s surgical textbook on Dixon's bookshelf. Chloroform came from the sixth-form lab at the Leys where it was used to kill the rodents we dissected for the then syllabus, no "namby-pamby" concern then for the rats or for squeamish or ethically bothered students.

The London Zoo (Dixon was a vice-president) brought similar privilege and a different animal interface. On a zoo visit – Sunday mornings, only open for fellows of the Zoological Society – we could, sometimes with friends, cuddle a boa constrictor or ride on the giant tortoise before climbing up the stairs to the restaurant where fellows and their guests had Sunday lunch. On occasion this could be scrambled ostrich egg. 'Fed 30, delicious,' pronounced Dixon.

Oxbridge may have been coined as a word by Thackeray, but that elitist hybrid goal was not, in our time, the feature of sixth-form competition it has become; certainly not for us.

Cambridge offspring went to fathers' colleges almost as of right. David Godwin, John Taylor, Laurence Reddaway, and Graham de Freitas were, in sequence, closest companions to John, Robert, Stephen and Richard. All were Clare fellows' children so all went to Clare as did the oldest three of us (though not actually a fellow, de Freitas senior was almost and became honorary fellow). Fathers arranged matters:

> 24*th* *June 1953, JD Boyd to Sir Henry Thirkill M.A., C.B.E., M.C.*
> *My dear Master*
> *I wonder if you think the time is appropriate to enter Robert for Clare? He is now 15 (May 14th) and will be taking his General Certificate next spring. It is not yet obvious what he will read when [not if] he comes up – but it will certainly be on the science side – possibly even medicine. I wonder further, if you think you ought to see John, who is now 17.*

> *My dear John,*
> *To let you know that I wrote to Sir Henry Thirkill some days ago about your having a talk with him.*
> [Dixon to John, 28 June 1953]

John's proposed admission as member of a Clare family was no surprise. The master was recalled by Dixon 'dangling you on his knees when you were <u>very</u> small and I remembered your being so interested in his watch which chimes. He told us the watch had been given to him by his batman after the first war world war – and his eyes filled with tears. There is some story there'. (Fob watches were kept in a waistcoat pocket, on a chain. Lifted out they could be clicked open to amuse the young. Robert remembers Sir Joseph Barcroft doing the same for him). In due course, the deed was consummated in a handwritten note

> 14th June 1956 *from the Master (H.Thirkill), the Masters Lodge, Clare College, Cambridge telephone Cambridge 3454.*
> *My dear Dixon, I have received your letter enclosing cheque value £13 on payment of initial fees for John and Robert. A receipt is enclosed herewith.*

This pen-and-ink conduct of administration by the master himself is of its time, and apparently not untypical of an institutional head. In the suitcase was an endorsed cheque for John's termly prep-school fee at the Hall School. It bears the personal signature of Mr Wathen, the school's headmaster.

The privileged track is also of its time. Five years later, when it was Stephen's turn, university expansion was the topic of the day as was more egalitarian entrance. At Clare, Bill Wedderburn – leftist lawyer and, later, Labour peer – led the reformist charge but, not without criticism.

> *For the Blythe feast at Clare I had as my guests James Bartley and Kingsley Amis (who, you may not know, has come to Cambridge as a fellow of Peterhouse [from Swansea where he and James were together in the English Department]. This was a pretty powerful group alcoholically speaking but everything went fine until Bill Wedderburn after dinner over drinks in the Combination Room tackled Kingsley over the latter's views which Bill regards as reactionary upon University expansion ... The problem [suitability of arts degrees for the proposed expanded entry], of course, is a real one. But it's discussion in a Combination Room after well lubricated feast by such protagonists was a WOW. I have not yet lost the reflected glory of producing someone who had the courage to tell Bill to his face that he was more than an egregious ass – he was a bloody fool. And, dear John, as I regard Bill as the source of the reform that has made life difficult for Stephen, the situation was not without pleasure for your father.*
> [Dixon to John at Yale, 17 December 1961]

Less in tune with the evolving emphasis on exam success, Stephen did get into Clare but the process involved senior tutor rather than master with nail-biting uncertainly replacing the easy assumption of five years before. Richard, in turn, opted for distance and went to the "other place" where Merton became something of an Oxford choice for children of Cambridge professors who preferred student life to be at a distance from home.

The apparently effortless easing of mature cygnets into the main river, was not, in reality, so effortless. Expected future "membership" of Clare had been underpinned by vigorous educational paddling over many years. Prep and public schools were chosen carefully, and extra progress encouraged. Long-term strategy was always in mind.

Thank you very much indeed for your letter of 30 April relating to Robert's work when he joins the sixth form next September. My desire that Robert should carry on some mathematics is conditioned by two considerations. Firstly, though he will probably do something on the biological side, and possibly medicine, there is no firm decision on this at the moment, and it may well be that he would like to do some physics, at any rate, in his Tripos. If he was to stop mathematics now it might be very difficult for him to make up his mathematics at a later stage. Secondly, if he does do biology, or even medicine, I think it will be most advantageous if he has a little mathematical training beyond that required for General Certificate ... quantitative aspects of biology are now so paramount.
[Dixon to 'J. Stirland esq., Senior master, The Leys School, 5 May 1954]

For John, then expecting to become a zoologist, and wondering whether to go on Scout camp, 'the major thing is to keep the ultimate aim in mind – and it might well be that you will gain in biological knowledge if you go to Scotland'.

"Coaching" was pervasive and our tutors superior. With Hansi Phillipsohn, refugee, former headmistress (and sister-in-law of physicist, Ewald), we learned to recite the *Lorelei* and other poems of the German canon. Mr Arnold, prep-school headmaster, improved Robert's maths – 'Thank you for your letter and cheque for £5.10.0.' – that, when teaching, he slid his hand up Robert's shorts was merely a mildly improper private frisson; perhaps for them both. A go-ahead son of EV Rieu, the translator of *Homer* for Penguin Classics and founding editor of the series, had introduced innovative colour-coding to help remembering Latin genders. He tried to improve Robert's progress with that language. Headmaster Butters of the Choir School (also of doubtful morals!) was planned as a coach for Stephen. To polish our fluency, all of us had French "exchanges".

Boyds were not unique in this careful choice of tutors. Other Clare families also sought the best coaches. John Taylor was helped with his maths by a junior PhD at King's who left to teach at Eton (probably Norman Routledge whose pupil, Timothy Gowers, went on to win the Fields Medal). His flute teacher solo-ed in the London Philharmonic. By contrast, the "writing off" of children in our circle who were not academic was very apparent. When another fellow's son failed to exhibit academic potential, we heard Amélie comment sympathetically that 'at least they will have fun with the younger two'. "Son" is relevant; the same father reminded Dixon that he had married Penny for beauty, not for brains. The bar could be high. Both Geoffrey Keynes, whose dynasty we will come to, and Biddy Barcroft had, in comparison with their siblings, considered themselves suitable to be written off.

Continued support did not end with graduation. Useful contacts in public life were pressed on John, postgraduate at Yale; early step to success at the Foreign Office. Professorial and more junior colleagues were recruited to prepare Robert for a post-Cambridge scholarship to his clinical school; first rung on the medical ladder.

5 Family Life

i   Help and Childcare

Servant was not a word used by our parents, but they had quite a few. "Help" was becoming the preferred word for the servants who were integral members of the Cambridge families we knew.

In our family, their hiring, their duties, their occasional crises, and their not infrequent evolution into becoming family friends were matters for Amélie. She provided the domestic skeleton, they the muscle in three key domains: childcare, housework and gardening. In keeping with her pre-marital aspiration, Amélie did much of the cooking. She became quite good. Dixon occasionally dried the washed-up dishes as we always did. Our other duty was to make our own beds, a mildly more laborious task in those pre-duvet days.

Most important to us children were the three who cared for us: Nanny, Emily Brignell; Hedi, later Hedy; and Norah, Norah Ansell. These very different childless women came more or less in sequence. All three wound through our lives over five decades and we through theirs. As time flowed, relationships evolved, both comfortably and less so. Nanny loved us. Norah loved Amélie. John and Robert and Stephen loved Nanny. Richard came to hate Norah, but Stephen was neutral about her. Hedy was loving and Stephen loved her, but she was resentful; Hitler's fault, we thought, as he had expelled her from Vienna. At the time they were to us extensions of Mummy, part of the family.

Nanny's Cambridge home where she lived with her elderly mother was, in both senses, the other side of the tracks. It was over the humped Mill Road railway bridge at 107 Cavendish Road. Lit by gas, with a coke range, it had a copper for the wash and an outside privy. Driven out of London by the doodlebugs or in the years before and after we had no doubt of her love. Though only a toddler, Stephen can remember 'bathing in a tub in front of the range'. It was a cosy experience. Stephen was gifted by her a loved piece of her furniture. John even named his eldest daughter,

Emily, in her honour though, usefully, he could also present the name as anglicised Amélie. Robert, in later life as a middle-class paediatrician, found himself drawing deeply on the warm experience of working-class life with Nanny and her mother in identifying with his inner-city patients. Friends there were class-different. At the Perne Road recreation ground, "the rec", they showed him there was no need to be pushed to make a swing swoop, you only had to use your legs. Paddling in the ditch nearby were leeches which an older girl kindly removed. On Cavendish Road itself we "played out" seeing how far we could walk along the war-emergency water-pipe without losing our balance. Providing a home at a time when flying bombs were falling on London probably saved Robert and Stephen's lives. To us, Nanny was very special. Lovingly anxious on our behalf – she wouldn't let us pick blackberries in case they were deadly nightshade – and always a reassuring presence.

Before the Boyds, she had worked for Dr Richards in Grange Road looking after Andy and Barry (note not "Mrs"; the husband labelled the family). These predecessors felt to us almost half-siblings; mysterious ghostly presences. Nanny had slept in Andy and Barry's bedroom as she did when she accompanied John and Robert to pre-war Belfast. Amélie's expectations were perhaps different for Nanny didn't share a bedroom with John and Robert in Grantchester Road; perhaps she even came in by the day on her "cycle".

In 1937, accompanying Amélie and toddler-John to Belfast, pre-Robert and pre-war, 'Nanny was rather homesick' but soon 'is happier'. Amélie, in typically Amélie fashion, went on July 12th (the day Ulster Protestants celebrate anti-Catholic victory on the Boyne), 'to Lisburn Rd. to show Nanny where the Procession was … in its elemental, childish delight and bigotry! The comic thing was our Catholic cook told N[anny] she must on no account miss seeing it!'. This was typical of Lennoxvale's non-sectarianism and of a happier Belfast in the years between the bloodshed of the early '20s and "the Troubles" of the century's end.

Robert was shocked when, as a teenager, he learned that Nanny had been paid for having him and Stephen to live with her.

He had assumed she did it for love. He is even more shocked, indeed horrified, to learn now from the letters, 75 years on, how Amélie schemed to get rid of this, to her, irritating children's nurse. 'Nanny is a trial but, for her, doing her best'. Or 'John [has] a slight cough and Nanny's dumbness accounts I think for this'. When war broke out and Amélie moved to Aileen's County Armagh general practice, she wrote to Dixon of regret that 'Nanny came along [as] she refused a berth for Cambridge'. A month later, she is pleased to tell him that 'Nanny's departure [to see her uncle who was dying in Cambridge] has been nearly all gain. The children have taken it easily. She ... write[s] me emotional screeds but I will give no word of her return, deathbed or no deathbed'. Nanny did not return but remained involved with Amélie and was used – Amélie again – to store furniture from Fernside (the spelling and grammar are Nanny's).

> *My dearest Lady,*
> *Many thanks for your very nice long letter...do not worrie about this furniture, we are only to pleased to be able to help you a little bit, there is not one thing in our way, We only wish you could see how nice the men have packed it in our front bed room, and they have got all in there, you know dear we never use that room, so it can all stay there untill you want it again. So please dear do not worry any more (will you). If we have any friends to stay we have other bed rooms. But we would just like Dr Boyd to come and have a look to see how nice the men have packed it. Now the Pram is quite alright, I cleaned it all up and have covered it right over, and all will be taken great care of, and I have washed up all I found about. You know dear one must take great care of furniture today, and not part with any, because after the war it will be very dear and very scarce. This is just a hurried note dear, as I want you to get it. I often wonder what the darlings are doing and wish I could just pop in to be with you all again, but still we hope that time will come again, and lets hope this may be a Victorious New Year ... now dear must run to the post with very best wishes to your dear self from mother and lots of love to you all and heaps of kisses to the darlings. Yours very affectionately*
> *Nannie [(sic) – is the different spelling a matter of class?]*
> *P.S. please write soon*
> [Emily Brignell 107 Cavendish Road, Cambridge to Amélie in Ireland, 6 January 1941]

While this may, indeed, be an emotional screed, Amélie's comments to Dixon [9 January and 5 March 1941] are unattractive. 'My only anxiety there now is that if we can get our furniture back from there will she consider herself included! ... I don't propose taking Nanny back – to the pictures I'll take her, yes, but nowhere else!'. But she did. By the autumn, JMcC comments [12 September 1941] that 'Your "Cambridge Nanny" will rejoice to have achieved her purpose and to be with her charges again, and I hope the arrangement will prove satisfactory and leave you some free time, tho' bye and bye you will likely wish to have someone more of the educated "mother's help" type – perhaps an Austrian refugee woman of the better class, of whom there must be a number who would be glad of employment and a congenial home'.

Before Nanny's return, this suggestion had been followed. Hedi, unlike many of her relatives, had escaped from Nazi Vienna and had ended up in Belfast where, initially, she worked for the Barcrofts and is, for the first time, mentioned in the letters. Amélie is busy in general practice and John and Robert are an issue. 'Biddy has generously and sincerely offered to care for them. Hedi is, too, most reliable and kind'. Later it is Amélie's turn, and she has both sets of children. The Newcastle house is small but

> *with Hedi in dining r[oom] and our meals in sitting room its*
> *remarkable how well we manage such a crowd.*
> [Amélie to Dixon, 10 April 1941]

Robert remembers that there was also Mary who made the beds, perhaps she came by the day.

Life in Newcastle settled and Amélie's cinema companion was no longer Nanny. 'I've just come back from *Pride and Prejudice* with Hedi'. That happy arrangement didn't last. Through renewed fear of German invasion via Ireland, or perhaps just through the catch-up of bureaucracy, control of where refugees could live was tightened and our lives were, to Amélie's fury, again disrupted.

Hedi's permit has been refused. Its damned stupid. Co Antrim police are rational but Co. Down are hidebound idiots. Shall be alone after next Mon. but will keep John Barcroft for company for John as he will be too much with me in any case once Hedi goes.
[30 May 1941]

Two five-year olds being easier than one sounds optimistic.

As we and our parents zigzagged between Belfast and Newcastle, Cambridge and London during those war years and after, childcare did too. Hedi/Hedy was in Ireland and, much later Cambridge. Nanny was Cambridge and sometimes in Ireland but never again London. In London and later Cambridge, Norah became the carer for Stephen and Richard. When Richard was two, Amélie became ill with "jaundice" and Richard went briefly to stay with Norah's family but 'was not on top form at the Ansells'. Norah herself became ill and Dixon felt 'awfully sorry about it but, at the same time, can't help but remember how almost nasty she has been with the kids. However, she has not been well'. It was crisis time; Nanny came, again, to the rescue.

Rickie restive when [Aunt] Helen left. Imagine him clinging to her! Relativity is the root of everything! ... we really are very deeply in [Nanny's] debt and must always remember it. Everything is done so ardently and with such grace.
[Dixon in London to Amélie in Belfast, probably April 1947]

Adjacent sentences including 'relativity' and [biblical] 'grace', are very Dixon. Is he also distancing himself from the unattractive attitude to Nanny expressed in his wife's letters?

A couple of months later Amélie has recovered, and Norah is back. Norah could indeed show a nasty streak which fostered Richard's antagonism and Robert can remember occasional physical or emotional violence but, when Amélie and Dixon went to Switzerland that summer for a convalescent holiday taking the older two, Norah was left in charge of Stephen and Richard, and Joan's Sarah, and took all three away to Swanage.

The "letting go" of nannies and carers as children grew up was always a problem for the middle-class. Were they servants or indeed members of the family? When Richard was five and less in need of care, the birth of Peggy's Michelle, the new cousin, provided a perfect opportunity. Norah was seconded to go to Lennoxvale with the baby where JMcC, now 86, was happy to welcome her.

> *Fine that Pegs and infant thrive. If P[eggy] is determined to come I cannot stop her and I need not say it will be a joy to have her company. If Norah is taking willingly and cheerfully to her new task, and if she is as good as of old, I feel P and the baby could not be in better hands. Tell Norah I shall be glad to see her again and hope she will be happy and comfortable while here.*
> [JMcC to Amélie, 21 February 1950]

Norah only briefly looked after Michelle. She then went to sea as "children's nurse" on the Union Castle Line. Amélie, typically, had kept in touch with Hedy who thus succeeded Norah in looking after our young cousin.

The roles of Nanny, Hedy and Norah evolved as the children grew but did not entirely disappear. Nanny continued to come in and out at St Chad's until she inherited property in Folkestone; on the front no less. Robert stayed with her to revise for his finals. She died before Amélie and, perhaps in keeping with earlier negativity, Amélie forgot to tell us; "letting go" indeed. We never mourned Nanny as she deserved.

Hedy, back as extended family, was left in charge when Richard was ten, 'most kind and good in my absence'. Marion Hines asks Amélie to remember me 'to each of your boys and to Hedy'. Amélie tells John of a weekend visiting Stephen at Bedales which was 'happy and busy and gay as usual. I took Hedy'. This was 30 years after Newcastle. 'We stayed in the Railway Hotel and were given enormous breakfast'. Stephen's emphasis is also on food.

> *Mummy and Hedy came down last week-end, the latter clutching her traditional tin of ginger buns, and we had a fine time out on the*

140

downs in July type weather, quarrelling about my allowance etc. and
cooking steaks.
[to John, 3 March 1961]

Back in Newton Road,

Hedy [is] recovering from a week of flu – which slows her down quite
considerably – horrid bug. But today she did football pools and had
bath so life is improving for Easter day. We feel a very small family
with you and Rob missing ... Even Hedy has left for the weekend, to
see her sister and a ballet.
[Stephen to John, 2 April 1961]

As with Nanny, there can also be parental irritation. At Christmas, two years later, Stephen comments to John that, 'Hedy [is] in a comparatively amenable mood (she is v. bitter now – Daddy is v. courteous to her and after she is in bed says it is time to give her a boot in the pants)'. Shortly after, leave she did, to share with a female dentist in Lensfield Road and to become her assistant. She then rather faded from family concern. Only Stephen remained warm. He remembers visiting her at her flat round the corner from Bateman Street.

Norah, returning from her Union Castle years, was drawn back in too. Dixon reports to Amélie, away on holiday with Peggy, that William Hamilton stayed and 'Norah catered very well – chicken last night'. Robert adds that 'Rick is being an angel and Norah is coping and, [dating the exchange] England has won the world cup'. Norah continued in and out tidying, gardening and grandchild-caring into Amélie's old age. Was she paid? She helped Amélie clear the house for her final departure, Cambridge to Oxford. We always considered Norah to be essentially a Boyd. But, for her family, the perspective was different. Her nephew's funeral address barely mentioned the Boyds; a salutary reminder of boundaries.

At Bentley Rd, before the war, housework was done by Violet, a local girl who was pleased to learn that Amélie did not expect her to wear a uniform cap as Sarah, at Lennoxvale, did save when she accompanied us outside the house. Nanny would

141

occasionally report on Violet's wellbeing over the decades which followed. Settling back in Cambridge, mid-war, an "Irish girl" would be needed. Recruitment, with assistance from Peggy, having been successful, JMcC gives some detail to son-in-law Dixon; Stephen was then barely two weeks old.

> *[I do] not think you need fear that Kathleen will be tempted to leave you. If one may judge from her previous record of 10 years in one place she is a 'stayer'. Being herself one of a family of nine, and having been in a household where there were, I think, 15, she is accustomed to children. After all, there are not so many households in England in which she will be treated in friendly fashion and not as a mere machine. Wages can gradually be raised. Helen, who spent part of her time here in searching for 'help' for various maidless friends, will have told you that people are offering 35/- to 40/- per week to induce girls to cross from Ireland.*
> [7 March 1943]

But there could be setbacks. JMcC is sorry to learn that

> *you are having "a Kathleen crisis" ... greatly concerned and extremely sorry on all counts. What is the matter? Has K. got in to trouble [become pregnant] with one of her numerous boy friends?*
> [to Amélie, 5 November 1943]

That crisis resolved, Kathleen stayed and on our return to London came with us. She was in the house when the doodlebug hit; probably also in the Morrison shelter with Amélie and Dixon. We can remember Amélie, once she acquired a post-war car, driving Kathleen, and her equally Irish successor, Bridget, maybe a sister or cousin, to St Joseph's on Sundays in an attempt, perhaps, to pre-empt 'trouble'. Memory is of once being told that they had both, after leaving us, become ill, probably with TB, the Irish malady. At the time we knew nothing of that.

In the '50s, "Irish girls" became replaced by those from either side of liberated Europe who wished to improve their English. They were seen as more equal to us than the Irish. Nia, older than a "girl" and life-long family friend-to-be, was Dutch.

She was still writing to her special love, Stephen, after he settled in Japan. Ruth and other *au pairs* were German.

> *Saw Emelyn [sic] Williams' "Accolade" at the Arts ... Quite original and interesting and good acting. Took Ruth – who is growing up and full of intriguing mind – I hope she has shaken off her Nazi shell by her English experience.*
> [Amélie to John, 10 May 1952]

Still later, *au pairs* no longer a feature, three local part-time dailies, Mrs Andrews, Mrs Kember and Vera (we puzzle why not "Mrs Stephens") did the housework. We were sometimes drawn in as supporting cast.

> *Mummy has engaged me to write ... to Mrs Andrews, who won't appreciate it, but has a stomach ulcer.*
> [Stephen to John at Yale, 17 April 1961]

Vera, who added cleaning for Peggy to her week was more feisty. Her description of a Bishop's visit to her hundredth birthday party is in character – 'I will always remember [it]!'

Of gardening, there is little written; surprising as Amélie was keen on her garden and often mentions it as a source of joy. Neither gardener, Arthur Stallybrass, nor Norah in gardening mode – in later years she spent hours in the Newton Road flower beds – receive a serious garden mention. Absence of comment on Arthur's tragedy, a son dying of muscular dystrophy, is also a surprise. Richard remembers Amélie's support and solicitude.

Washing machines or carpet sweepers, unlike those who used them don't figure much in the letters. In memory they do. At Nanny's, we cooked on the range, fuelled with coke from the store at the far end of the yard. We lived in the kitchen, warmed by it; the front room was never used. At Boyd houses it was gas cooker; only our more sophisticated friends had electric ones. Coal gas, piped from the gantried Cambridge gasometer, had its smell, worrying if there was delay with the match. Lighting at Nanny's was gas too. She had no electricity, and upstairs, no gas either. In the kitchen, being lifted up at dusk to put the lighted spill to mantle – a scary pop, and the rewarding flush of light – was a pre-bedtime treat.

The "accumulator" from the war-time *wireless* at 107, had to be charged at the Mill Road shop; a pleasant little walk with Nanny. In other houses we knew, lights and wirelesses were on *the mains*. Electric shocks were a feature of life and could be severe. 'Never touch the switch when you have wet hands boy, or with wet feet. You might die'. Fuses blowing were a recurrent challenge, always dealt with by Amélie. Inserting the new fuse wire (correct thickness varied with the amperage) was, like driving and mending bike punctures, a female role in our household. Elsewhere, friends' fathers drove the car when there was one. Dixon did not. (Cars could not be bought during the war, and then, into the '50s, only after a long wait on "the list". We were no exception. Amélie rode a 98cc motorised "autocycle" to get to her clinics).

Bathing at Nanny's, after an enjoyable afternoon playing out, was in the galvanised tub, Stephen remembers. The Boyd homes had white, if flaking, baths in bathrooms. In earlier days unreliable hot water came from a gas geyser, the "Ascot". Its pipe opened above the bath to deliver either scalding or, when it blew out, horribly cold water. At Nanny's, the water closet – WC – was in its whitewashed outhouse. Squares of newspaper hanging from a loop of string were scratchy wipes. The spiders were scary. In those times a grown-up had to undo the braces' buttons at the back of smaller boys' shorts so that we could "go"; and then reverse the process.

At Nanny's, washing was done – on Mondays – with a fire under the copper. Clothes and sheets went through her mangle, fun to turn the handle, before ironing with flat irons. While one of the pair was in use the other heated on the range. The first washing machine in a Boyd household arrived around 1947. It had an electric mangle between whose powered rollers Robert was nearly ruined. The rubberised maw pulled in his exploring fingers and they wouldn't stop – only John's leap to throw the switch saved him. Before the machine, Boyd sheets went out to the laundry, but woollens were washed by hand and, to dry straight, laid under newspaper on the floor.

Through the '40s, clothes-rationing dominated. Our shirts and shorts were usually "cast-offs"; from elder brothers certainly, but also from slightly older children of friends or colleagues. The name of Gabriel sewn into them became familiar; we never met the Gabriel children in real life. Their father was Dixon's colleague at the London.

Daily walks to the shops and frequent deliveries were relied on into the '50s. Milk came in bottles, by horse and cart – to Stephen, those placid animals were enormous. The groceries came on a bicycle. Garden produce was stored for the winter, especially after moving to a larger house and garden. Apples on racks on newspaper; Carrots clamped in sand. Without a fridge, food only lasted a few days unless bottled, made into jam or, for eggs, smeared with "waterglass" preservative, to keep them for baking. For breakfast eggs (cooked breakfasts were deemed essential) we kept chickens, as did Nanny. Children had, at Nanny's and at home, the duty of feeding and the slightly scary pleasure of sliding a hand under a "layer" to feel for an egg. Food could involve brutality. If we were to eat chicken, a rare treat, Daddy had to wring the neck of one from the coop. On one occasion we surprised a fox, there for its own purposes. It was surrounded by corpses. Of shop eggs, bad ones were an appallingly smelly unpleasant surprise.

Acquisition, in the late '40s, of the Boyd's first fridge transformed all that. It was second-hand.

ii Politics

Dixon, in his League of Nations days was of the left; extreme in his case, less so for Amélie. Richard thinks he once saw a Communist Party membership card. If so, we cannot now find it. Amélie reported his declaration, at the time when marriage was first in question, that 'you mustn't mind if I am imprisoned for my beliefs'. At Queen's he met student radicals of Irish independence. MacConaill (a Walmsley anatomy protegee, in due course to

145

become a professor in the discipline), was one of them. The third generation of his family dedicated to independence, he is recorded as being a "hero of the Republican Movement" for his role, while Queen's University medical student, in organising first aid for Irish Republican Army fighters during the Battle of Belfast. Subsequently, he organised British first aid in Sheffield during its German Blitz – portable skills! John Doggart, engineering student, regaled us with his role in Ulster gun-running. Sara's husband, he became a much-loved family friend. His successful company (manufactures included *hula hoops*) eventually floated on the stock exchange.

Most former students moved rightwards. Dixon certainly did, Amélie, less so. A letter from Ralph Meredith describing a Mosley meeting in the Albert Hall is of the earlier period.

> *Blackshirted stewards clustered around the doors. And a triple cordon of police outside, and beyond them a yelling crowd. And Oswald [Mosley] standing in a spotlight so bright you couldn't see his face, orated for an hour and a half. He gave them everything: cold (and not at all accurate) logic, passionate emotional protest against injustice, and hatred. He was best at the hatred, with a marvellous sneer in his voice that seemed to cut through the audience like a knife ... and the audience rose best when the hatred was turned against The Jews ... 'We fought Germany once in British Interests. We are not going to fight her again in Jewish interests' and this time they nearly brought the roof down ... this kind of thing is dangerous especially as it has the support of the police, and apparently of the magistrates. There is a point where Free Speech becomes anti-social license.*
> [Ralph Meredith in Chelmsford to Dixon, Carnegie Institute, Baltimore, 4 October 1934]

Ralph, responding in another letter [26 December 1934] to an apparent Dixon-comparison of Soviet and Nazi policies, seems, even then, more left than Dixon. 'Take care that it is because you genuinely think that they are worse, and not merely because they are 'new'. There's Spain, Manchuria, Bulgaria, South Africa etc ... Think of Russia think of collective farms, think of Germany think of concentration camps. That is roughly my attitude'.

The letter goes on to update Dixon on mutual friends, Bartley, Calvert and Rodgers, and on the pro-working-class attitude of, new QUB vice-chancellor, Ogilvie.

Despite then-closeness they were diverging politically, and socially. He never came to stay in our time and Dixon's mention of a wedding present for Ralph, is downbeat. 'In response to the letter and cheque I sent have heard nothing from Ralph. But I suppose he doesn't take these matters very seriously'. (Amélie, widowed, was a great believer in picking up old threads. The next letter from Ralph to survive is to her half a century later; no real surprise).

Come the war, Mosley is in detention and Amélie expresses to Dixon, as we have already seen, pretty decisive scorn for Ralph's alternative. Dixon merely wonders 'is it adult or simply escapist to have lost one's early idealistic hopes?'. A decade on there was a general election. It is 1951. Amélie reports Margaret Hill's advice as being 'to hold one's nose and vote Tory'. Dixon, more stridently, celebrates that the 'Labour bastards' are out. Some shift! We only heard his repeated criticisms of Russian imperialism as groundless teasing of the young – if only. At the personal level he continued to support positions that were then radical; humanism and birth control and opportunities for those from (new phrase) the third world. Malcolm Potts, his former PhD student, and later an international leader in population management writes [to Robert, 28 March 1994] that 'I have two photos on my bookshelf and one is of Prof Boyd ... Your father not only inspired my interest in embryology, but he supported my excursions into the politics of abortion – which in the 1960s was a revolutionary and potentially embarrassing thing'.

iii   Jews, Blacks, and Irish

Six million dead Jews, skin-colour racism, Irish religious strife – civil war and partition barely a decade before John's birth – were absolutely not domestic issues for us children. We saw ourselves as being of a liberal English family. Jewish colleagues and relatives, African and Asian postgraduates, Catholic friends

147

wove, calm and composed, in and out of our house and our lives. All were educated middle-class, perhaps a relevant point. All were treated with overt respect.

Words sometimes used in the letters or in conversation and which now shock – 'yid'; 'little jew'; 'tarbrish'; 'picaninny'; 'breed like rabbits' (of Ulster Catholics) – are not antisemitic or racist when in context, with the possible exception of the rabbits. But such words do indicate an unwillingness to take exception to the casual assumptions of the time; a preference to be in the English tent rather than outside it and an acceptance of associated privilege. England's role as a beacon of enlightenment with a justified international status (which was slowly being eroded) was taken by us as self-evident, probably by Dixon and Amélie too. The possibility that England's position in the world stemmed from slavery and pillage was not yet part of public or, for us, private discourse. A hint of English superiority towards Americans and white "Colonials" was also occasionally apparent. In short, we were a normal mainstream mildly eccentric upper-middle-class liberal family. We didn't go to church and perhaps swore more freely than most. Dixon, unlike most fathers, did not drive, or mend our bicycles.

There is no hint in letters or in memory that our parents felt at increased war-time hazard through Amélie's Jewish origin. Proud Iklé descendants, but called Boyd, we absolutely didn't consider ourselves part of any Jewish community and viewed Jews and Jewishness from the non-Jewish mainstream; a mainstream which was sometimes antisemitic.

> *Spooner [Clare fellow and bacteriologist] is definitely reactionary and, I fear me anti-semitic – the bloody fool – his subject owes more to the Jews than almost any other, Koch, Ehrlich, Waldeyer, Cohnheim etc etc. He has a stupid mother, however, and there is something in heredity.*
> [Dixon in Cambridge to Amélie, Lennoxvale, Belfast 1936]

Many of our closest friends, at least in London, were Jews, albeit secular ones. But, at the time, we did not think of labelling them as belonging to that category. Being ourselves called Boyd, rather than Loewenthal, John and Robert's prep school headmaster could casually mention to Amélie that the school restricted its Jewish entry to a quota. Uckos, Sonneborns and Levines, all in Robert's class and welcome friends, would not be allowed to become the majority. We doubt if Amélie mentioned our tacit contribution to that quota.

Karl Pearson, the founder of biomathematics, spoke of applied eugenics at a dinner in his honour in 1934. Eugenics 'culminated in Galton's preaching. Did I say culmination? No that lies rather in the future, perhaps with Reichkanzler Hitler and his proposals to regenerate the German people. In Germany a vast experiment in in hand, and some of you may live to see its results. If it fails it will not be for want of enthusiasm, but rather because the Germans are only just starting the study of mathematical statistics in the modern sense'. The slender volume recording the occasion was amongst Dixon's books. Less than ten years later. during the rather different culmination which actually took place, JMcC asks Dixon [7 March 1943] 'what sort of punishment, drastic and lasting, could you suggest for Hitler, Goering, Goebbels and co. if captured alive? I would like to leave it to the Jews, but that they might make too quick an end'. He doesn't seem, when writing thus, to remember that his daughters belonged to that very group (it is notable that victory is being assumed long before the landings of D-Day). Even earlier, by Christmas 1942, he understood the reality of the concentration camps. 'To share our turkey ... [I] roped in two of Peggy's lonely refugees ... as a good deed! They have had hard times, but at least they are fortunate to be here and not in a Polish ghetto or concentration camp, where they fear some of their relations are, if still alive'. Peggy was indeed a vigorous supporter of refugees in pre-war Belfast. So were Dixon and Amélie, somewhat less vigorously, in Cambridge.

Though the Boyds and Loewenthals opted out from Jewish identity in Britain, their relative lack of response to events in Germany and scant comment on them afterwards does surprise us. In no letter after the war do the words Holocaust or Auschwitz appear. Peggy had left Belfast to work in Belsen, the first camp "opened" by British forces where, said Richard Dimbleby of the

BBC, 'lay dead and dying people. You could not see which was which ... The living lay with their heads against the corpses and around them moved the awful, ghostly procession of emaciated, aimless people'. Amélie never comments on Peggy's work there and JMcC's sole remark concerns a postcard from Peggy to Margaret, the Lennoxvale cook. It is dated from Belsen a year to the day after that broadcast. She 'says she is going to Berlin for 2 days! I don't like the idea of her roaming about in Germany so much on her own and Berlin cannot be pleasant now'. No letter we have mentions the camp although, on leave from Belsen, Peggy did give a talk to the Belfast Rotary Club:

> *There are two types of people now; those who have been in a*
> *concentration camp and those who have not. This remark was made*
> *to me by a Rumanian girl cellist who had been forced to play*
> *classical music in an orchestra outside the gas chamber in one of the*
> *Nazi extermination camps.*
> [Fink M.G. *From Belfast to Belsen and Beyond*, 2008]

Of all the letters, only a single one refers even obliquely to the Holocaust. It is from Amélie's Aunt Amélie who left Germany just in time and escaped from Portugal to the US in 1941. Her Waldeck son-in-law, Amsterdam friend of Anne Frank's father, was taken off a French train and shot while trying to reach Portugal. Post-war, that Amélie updates on family activities. Widowed daughter Clara Waldeck, Boyd Amélie's cousin, will be going to Cuba to join up

> *with a Frankfurt school friend [who] ... was taken away by the Nazis*
> *and suffered a lot in concentration camps.*
> [Amélie Lewandowski to JMcC, 20 July 1947]

This is the only direct reference we can find. Fourteen years before that Cuban holiday and four months after Hitler's rise to power things were already getting bad for the Jews but our mother and Dixon were about to wed and hardly seem to notice. Cousin Clara writes from Frankfurt [to Amélie 19 May 1933], where her father is a banker, that 'Papi has a lot to do with Jewish things and nearly never comes home before 9 or 10 o'clock in the evening. His business is very quiet'. "Papi" sends niece-by-marriage Amélie his 'heartiest wishes for the future'. He does not mention the situation. His wife does [Amélie Lewandowski to Amélie 18 May 1933], apologising for delay because 'over here everything seemed to go topsy turvey (is that right?) ... We got accustomed to the antisimitic [sic] atmosphere and I feel already quite settled down again. Fritz [Lewandowski, Clara's brother, later to become Fred Lenway, successful US businessman] does not resent the movement at his school and I am glad of it ... We don't go out much for the moment. Even the cinemas have no attraction for us just now'.

There had been a plan for the two Amélies to meet up before the wedding but, with a retrospectively chilling lack of attention, Amélie junior, writing to Dixon [6 July 1933], refers to the situation in five casual words. 'Auntie A. could not leave. At Open Air Shakespeare Theatre last night. It was <u>lovely</u>'.

There is scant evidence of much practical action as things worsen. One month before war breaks out there is a letter, in German, to Amélie's father from a more remote cousin Clara [Clara Flatow to JMcC (in German), 5 July 1939]. He appears, late in the day, to be sponsoring her. '[W]e will praise you, good Jack, as my benefactor on the day when I can finally apply for my emigration passport'. She survived but in Sweden rather than Britain.

Margaret Hill's husband, AV Hill, played a key role in establishing the Academic Assistance Council to help 'Jews and other undesirables expelled from German universities'. Dixon joined its committee but probably not until later. He and Amélie were domestically supportive of refugee colleagues in Cambridge (many of whom were, in response to an early-war populist panic, interned as aliens in the Isle of Man). One was Jacobson, a distinguished histologist who, earlier

> came to see me partly about his being a British subject and partly because of the worry of having his little money confiscated as an alien. He wondered if you would mind putting about £100 in the bank or in safe deposit for him. I see no objection but you had better mention it to your father. His wife will send the money to you in some way or other. God what a mess it is for refugees, thrown out of Germany and now under suspicion here.
> [Dixon to Amélie, 27 September 1938]

Ten years after the war, Dixon, on a work trip, writes from Macy Foundation, New York to Amélie [8 March 1955] that 'Charlie [Uncle Charlie] sends warm love to you all … George [Lewandosky] is going to California for a month. Fred [his son, now Lenway] apparently flourishes … [as does] Fred Iklé [Judith's brother]. The Jacoby boy [Tony, son of Aunt Olga's successor] is in New York and is now in a job of some sort'. Four Iklé lineages, from our extended family alone, thus contributed to the vast migration of European Jews to the new Jerusalem; a migration not especially highlighted to us children.

The Boyd attitude to other "ethnic minorities" (anachronism again) is liberal of the day. Cambridge colleges had had Indian fellows since at least Ramanujan's election to Trinity in 1918, but Dixon told Richard he had been the first to take an Indian to the Cambridge Graduate Science Club. He recruited Commonwealth postgraduates with dark skins (Not then, "Black" or "Asian"), several of whom became later distinguished in Cambridge or further afield. As an example of their casual non-repudiation of current attitudes, Grillo could be both a clearly accepted junior member of the department on an upward track yet have his name attached to Dixon's dog by Neville Willmer. Robert, in his schoolboy naivety where the few black boys at the Leys were especially popular and where his best friend was half-Chinese, remembers listening to Dr Grillo talk of racist experiences and finding his views paranoid – again, if only!

Of Ireland, we wondered why its stamps had such a funny name. Amélie and Dixon, while appreciative of friends, of memories and, Amélie, of its countryside, often expressed anti-Irish, or at least anti-Catholic, horror at perceived obscurantism in the "south". They scorned dismissal of contraception. Catholics did 'breed like rabbits'. Consistent with that scorn, not one member of our wider family that, across three generations, prided polyglot fluency in French and German for all and Spanish, Portuguese, Italian, Russian, Chinese or Japanese for some, could speak even a few words of the Republic of Ireland's official tongue.

As teenagers we found the parents attitudes to Jews, members of ethnic minorities and to Ireland entirely unexceptionable. By contrast, we found Dixon's anti-Sovietism derisible.

iv Money

There were only a few financial papers in the suitcase. Stephen and Richard think money is boring but Robert finds those few fossils interesting. Sometimes they surprise.

Amélie was always money-careful. Into old age she would make coffee in a hotel room rather than have it on the dinner bill, buy a bottle of vermouth for pre-food tipple rather than use the bar. She encouraged us to be careful too. John and Robert were seven and five when pocket money was instituted, 3d weekly for Robert which sounds rather a lot; maybe 6d for John. It was on a war-time summer holiday in St Neot's. The initial payment was accompanied by a homily and a little notebook, 'write down what you spend'. This was a family trait. As a bachelor, JMcC had kept a similar notebook ('April 29 1902, Milligan, 19/-, 1 ton coal'). Robert was still, as a young husband, similarly recording expenses. As children, perusal of the record could lead to uncomfortable moments. Pre-teen John and Robert, voyaging unaccompanied to Belfast, ordered morning tea and biscuit in their cabin as Amélie, unusually self-indulgent in this regard, always did. We were rebuked for this unnecessary extravagance. Similar auditing of teenage income led to exposure of Robert's paedophile.

If these memories make Amélie appear skinflint that would be a wrong conclusion. In "sensible spending" she was generous and, while she was proud to travel third class rather than first and was always reluctant to take a taxi, plans for "a travel" would not be abandoned on financial grounds. Dixon was relaxed in funding ice creams and other indulgences perceived by her to be wasteful. Focussing on income and increasing it where possible was, for him, more important than saying no to us. If both parents were present, decisions on treats were, to our regret, left to her.

Amélie never pressed Dixon to seek the financial rewards of clinical work or encouraged him to make a career decision on financial grounds. Cambridge was no exception. The Chair offered, Dixon accepted it despite the 'loss of £400 per annum apart from extras'. (A few years later, Dixon was one of the Board of Electors for the Chair of a different discipline, pathology.

The first choice of the board declined the offer on financial grounds. He had five children to support. He subsequently became President of the Royal College of Pathologists and Richard's wedding reception, when he married one of those five, was at that college. Would a fifth child have tipped the balance against Cambridge for us?).

Income beyond Dixon's salary could be substantial. In 1942-3 (the only financial year with full details of a tax return) other earnings added more than a quarter to his professorial salary of £1100: £245 for examining at other institutions and £59 for various external lectures. The Clare £50 fee for "fire-watching" is included. By 1948-9, salary had risen steeply to £2575 and examining and lecture fees, together with royalties, probably from sales of *Human Embryology,* added another £500, Amélie's medical earnings a further £250.

Another extra is fun. Dixon occasionally told us of work on human remains for detectives from Scotland Yard. His claim was verified when Robert, with other Leys sixth-formers attended a public lecture of reminiscence given by Francis Camps, then a renowned Home Office pathologist, who told us of 'A most remarkable diagnosis made by a colleague. He is now Professor of Anatomy in this university. The Professor took one look at the body and said it had been thrown out of an aeroplane. Indeed, it had been'. Admiration from those sitting alongside was exquisite. Income of £21-0-0 for 'Report re Bone Fragments' provides documentary confirmation of this somewhat unusual professorial activity. Other small payments are echoes of Empire: 'Fee for attendance at a Selection Committee [held in London] for a university appointment in the Republic of Sudan'; 'appt board, lecturers University College Ibadan'; external examining fee, 'University of Capetown'; payments for examining PhD candidates with names like 'El Angoury'. The tax returns also provide lists of professional tax-exempt book-purchases and of subscriptions paid to Learned Societies. More domestically, '2 maidservants' are mentioned. Such description of Kathleen and Norah would have seemed to us bizarre.

Swapping tickets so as to travel more cheaply released an easy extra: 'Refund from TWA for replacing first class ticket with economy ticket £58-16-0'. As examiners were considered "first-class" as far as tickets on trains were concerned, Dixon probably travelled "third" and pocketed the difference. Two generations later, when MPs were found to do the same, such class-linked behaviour was (another, albeit minor, cultural shift) labelled an "expenses scandal".

Some nuggets shed light on family dynamics. Dixon's bank cheques (once "cleared" they were returned to account holder) include some endorsed with the signature of his half-brother, Kenny. Kenny had became a dropout, accused of pilfering from the Greenisland post office. Perhaps Dixon was giving him an ongoing allowance unevidenced elsewhere. Quarterly cheques to Amélie for £125 remind Robert that, on his marriage, she advised him, to do the same. 'Provide her with some financial independence'. Weekly £10 cheques made out to "self" are a pointer that banknotes came not from ATMs but across the counter, credit cards, Paypal and Google Pay then but science-fiction fantasies.

Of the Boyd houses, Bentley Road seems to have been rented and there is no mention of purchase for pre-war Fernside. In 1943, a year of Dixon's income could buy a house like our 18 Talbot Road. By May 1947 when they sold it for £2900, its value had 'more than doubled in spite of the bomb damage' and his income had doubled too. They replaced that house with a larger one costing, from memory, £3000. Back in Cambridge and after renting St Chad's, they eventually purchased 21 Newton Rd (three floors and a substantial garden) for, Richard thinks, £6000.

The Newton Road house was leasehold and buying out the lease from Trinity College 15 years later cost the same sum again. They remained there for the rest of their married life. By the time Amélie left 40 years later, and to the boys' ultimate benefit, Cambridge house prices were many multiples of a professorial stipend.

8. Moving house – *John's version of a change-of-address card, 1957: Four boys, dog Dinghy, and cat Whiskers; parents bringing up the rear.*

9. The Boyds at 21 Newton Road Cambridge – *from left: front Dixon, Amélie, Richard; behind John, Robert, Stephen. About 1958.*

6 Who Knew Who

i Scientists

For a university to be more than the sum of its parts its members must interact. In Gwen Raverat's childhood,

> *the regular round of formal dinner-parties was very important ... The guests were seated according to the Protocol, the Heads of Houses, by the dates of the foundation of their colleges, except that the Vice-Chancellor would come first of all. After the Masters came the Regius Professors in order of their subjects, Divinity first ...*
> [*Period Piece*, p78]

Fifty years later, hierarchy had become fluid; divinity professors replaced by prize-winners, leadership more international, informality common.

Our parents were not, amongst their contemporaries, serious high-flyers. Dixon had a Cambridge chair but no Nobel prize, not even Fellowship of the Royal Society. Amélie was not president of a royal college or even a consultant. She only rose to the status of becoming commissioner of the local Girl Guides and treasurer of the Cambridge branch of the Medical Women's Federation (Dixon's secretary kept the accounts). They may not have achieved the first rank of academic life but they mixed with many who did. Professional and social standing was less driven by academic stardom than one might suppose. Cambridge networking, though not described as such, was daily life for our parents, and for us.

For Dixon, colleagues, great and small, were sat next to at high table, run into at meetings or encountered when wheeling a bicycle across the Downing Street site where anatomy, physiology, biochemistry and cognate departments juxtaposed. For Amélie, never Dixon except at David's or Heffer's, shopping encounters were a commonplace. In Sainsbury's – no supermarket then – men in aprons weighed out sugar into blue bags, cut the cheddar, and, until 1954, snipped the ration book. She would run into friends in the little queues before each counter.

Listening to papers from colleagues and their guests at the Ray Club (biologists) or the Graduate Sciences Club (scientifically wider, Dixon became secretary) was a good way to be informed of developments. Discussion could expose ignorance in even the most distinguished, Dixon returned home, admiring but amused, when our landlord, elderly Lord Adrian, displayed confusion between genes and chromosomes.

Invitations "in college" to lunch or, in the evening to hall were important to academic life. So were formal college "feasts" or dinners (occasionally even including wives). Conversation with those sitting nearby enabled one to learn what people were "up to" in their subject. In that period, long before national attempts to assess research quality, chats at high table or in the interstices of college business allowed the academic standing of individuals both junior and senior to be informally and candidly, and sometimes prejudicially, discussed. Decisions on appointments, promotions and preferments would follow. A professor had to go out of his way to keep an ear and a nose alert to aspirations, talents, and activities.

We recall and the letters confirm something of the range of individuals Dixon and Amélie knew who, in those years, transformed the scientific or even the wider world. Seventeen Nobel Prize winners, mentioned in letters or remembered by us, are among those our parents mixed with.

One evening at a routine college dinner is worth a quote. It was in wartime during the build-up before El Alamein.

> *Last night I went to Caius with Munro Fox [comparative anatomist, like Dixon evacuee of London University] and was in most distinguished company – no less than two O.M.s! Sherrington and Adrian [joint Nobel 1932]. Sherrington charming and so alive and active-minded – and he is about 80 [actually 85]. I talked to him about [his] Goethe lecture (my luck to have read it) and then a lot about neurology with Adrian joining in. I was, for the first time, in mental contact with Adrian – he is so remote and curious that never before have I felt anywhere near – even when beside him. But some*

RAY CLUB

Saturday, 25th May 1957

ST JOHN'S COLLEGE COMBINATION ROOM

Let it not suffice us to be Book-learned, to read what others have written . . . but let us our selves examine things . . . and converse with Nature as well as Books, Let us not think that the bounds of Science are fixed like *Hercules* his Pillars, and inscribed with a *Ne plus ultra*.

JOHN RAY, 1691

MENU

Pâté en Terrine
Meursault 1953

—

Homard à la Newburg

—

Filet de bœuf Rossini
Pommes nouvelles
Haricots verts
Vosne Romanée 1949

—

Fraises Brillat Savarin

—

Canapés Diane

—

Dessert
Tuke Holdsworth 1927
Château Latour 1934

—

Café

10. Ray Club Dinner, 25 May 1957 – *note the wines.*

LIST OF MEMBERS, 1954—55

F. J. M. STRATTON, Caius
T. KNOX-SHAW, Sidney Sussex
H. THIRKILL, Clare
G. I. TAYLOR, Trinity
W. H. MILLS, Jesus
H. H. THOMAS, Downing
E. D. ADRIAN, Trinity
J. GRAY, King's
J. M. WORDIE, St. John's
W. B. R. KING, Magdalene
F. DEBENHAM, Caius
F. J. W. ROUGHTON, Trinity
F. I. ENGLEDOW, St. John's
R. G. W. NORRISH, Emmanuel
C. F. A. PANTIN, Trinity
A. N. DRURY, Trinity Hall
J. A. RATCLIFFE, Sidney Sussex
C. G. DARWIN, Trinity
C. E. TILLEY, Emmanuel
H. GODWIN, Clare
B. H. C. MATTHEWS, King's (Vice-President)

J. F. BAKER, Clare (President)
A. R. TODD, Christ's
A. L. HODGKIN, Trinity
B. C. BROWNE, Trinity
R. L. N. GREAVES, Caius (Secretary)
A. B. PIPPARD, Clare
C. W. OATLEY, Trinity
J. HARLEY-MASON, Corpus Christi
W. H. THORPE, Jesus
B. G. NEAL, Trinity Hall
F. G. YOUNG, Trinity Hall
A. F. HUXLEY, Trinity
J. D. BOYD, Clare
G. K. BATCHELOR, Trinity
S. V. PERRY, Trinity

Non-Resident :

V. J. WOOLLEY
T. R. ELLIOTT
A. V. HILL
R. E. PRIESTLEY
E. G. HOLMES
E. V. APPLETON
F. BALFOUR-BROWNE
H. HARTRIDGE
R. J. PUMPHREY
E. J. MASKELL
H. McCOMBIE
P. M. S. BLACKETT

E. K. RIDEAL
J. D. COCKCROFT
P. I. DEE
O. H. WANSBROUGH-JONES
E. T. C. SPOONER
E. C. BULLARD
D. G. CHRISTOPHERSON
G. B. B. M. SUTHERLAND
W. A. DEER
D. G. CATCHESIDE
J. T. SAUNDERS
J. E. LENNARD-JONES

1954-55

Saturday, October 30   F. J. M. STRATTON
" The Scale of the Universe "

Saturday, November 27   A. N. DRURY
" Stored Blood "

Saturday, January 22   J. GRAY
" The Life of a Spermatozoon "

Saturday, February 26   S. V. PERRY
" Motility and Metabolism "

Saturday, May 7   W. H. MILLS
" Cambridge chemists of the XVIII Century "

ANNUAL DINNER

11. Cambridge Graduate Sciences Club Programme
1954–5 – Membership is scientifically distinguished; Dixon later became its secretary.

good Burgundy and the longing [?] to live up to Sherrington put him, I suppose, on his mettle ... Sherrington, too, was very interesting on Body-Mind relationships. The reaction against Behaviourism in one who is more responsible than anyone else for our ideas on reflex action is unexpected, perhaps, but very much in the tradition of the really great neurologists. He is writing another book – a sequel to Man on his Nature and Munro Fox tells me later that he writes all morning and afternoon every day. Wonderful at 80 and with so many laurels to be contented with. A sense of humour too. He told us a story of the days when he was working on chimpanzees – he was leaving the lab one night and had just locked up the room in which the apes were kept. Then, it suddenly struck him "what do the apes do when we go off?" So, he tiptoed back to the door and peeped through the key hole. For a second or so he could not make out what was in his field of vision. Then he realised – it was one of the chimps doing just as he was! Then – when the

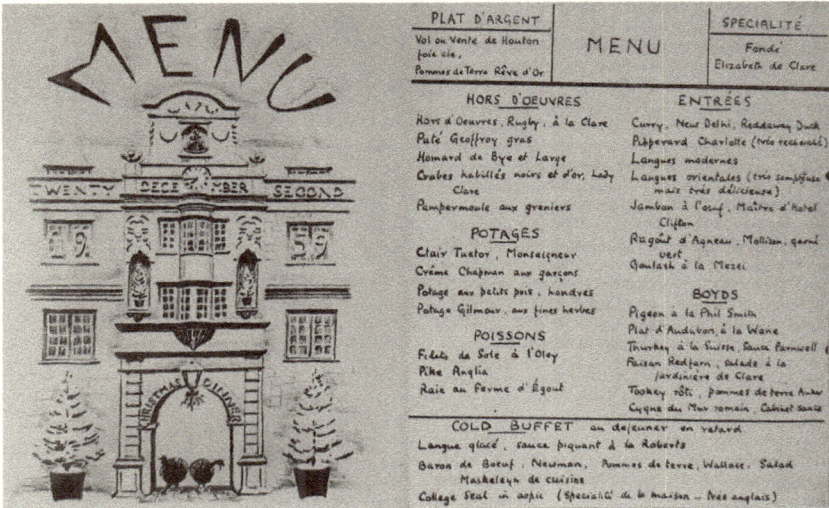

12. Christmas Dinner at Clare 1959 – as depicted by Fellow, Neville Willmer. Spot other Fellows' names: Reddaway, Pippard, Boyd etc. Wives were invited.

story had registered, he added coyly – of course it was the one female we had in the colony – "curiosity is, you know, more developed in that more reasonable sex!" It is so pleasant to meet humility at its best. Adrian had been in Baltimore on his recent American trip and spoke of Marion [Hines], Sarah Tower and the rest. I think he doesn't quite understand the American female scientist but he admitted their work was very good. Then, at about 10, Fox suggested we go around to Gray! [professor of zoology]. I, feeling several contretemps on my mind, tried to get out of it but couldn't gracefully so I went and the Grays were very charming and Gray is coming to see me about some anatomical problems. But he isn't a Sherrington and she is curious – but full of her two children. She asked nicely for you and about J[ohn] and R[obert]. [The Grays, like us, lived alongside King's Choir School. He later gave Richard his collection of bird eggs! The Gray children had been adopted].

[Dixon 26 Grantchester Road Cambridge to Amélie, probably in Ireland, 13 August 1942]

161

Another, more trivial report is of

> *a feast at Caius last night – and a good one, too, gastronomically and*
> *vinously ... sherry with soup, hock with the fish, burgundy with the*
> *meat etc – and it seems so appropriate – like ham and eggs ... At the*
> *dinner I met, inter alia, Sir Howard Florey who is now Professor of*
> *Pathology in Oxford and who I once met when he was Professor in*
> *Sheffield in the early "thirties". [was Dixon being considered for a*
> *job there or just visiting MacConaill?] I didn't know then, nor, I*
> *expect, did he, that he was going to achieve world fame and a Nobel*
> *Prize [1945] as the re-discoverer of penicillin.*
> [Dixon to John, 20 February 1954]

Amélie writing to John about a teenage cash crisis (the Swiss
money she had given him for a skiing holiday was out of date)
refers casually to another of the greats. 'Fearfully sorry! Will you
try and change it in Wengen and if not easily possible borrow
equivalent from Dr Sanger?' [a keen skier; Nobel twice, [1955,
1980]. Another contact is baby-domestic – John sharing bath-
time with Patience Bragg [Father and grandfather, joint Nobel
1915]. At teenage dances Robert partnered the daughter of Alex
Todd (neither much enjoyed the occasions). His wife was
daughter of Henry Dale [Nobel 1936] who, to Dixon's pleasure,
commented favourably on an article he and William had
contributed to *Nature*. Todd himself won the Nobel in 1957.
Michael Crick, a Bedales student slightly older than Stephen,
went to the same dances as Robert. He was collected from them
by his father, Francis [Nobel 1962]. Girls knew to watch out for
Francis. Belfast could be a link too.

> *Yesterday we had the Bethes to tea. She is an Ewald daughter, he*
> *[Nobel 1967, on sabbatical from Cornell with Caius fellow, Mott,*
> *Nobel 1977] is the atomic physicist, very much concerned with the*
> *initial Los Alamos affair; and a really charming man. His father was*
> *a distinguished physiologist in one of the German universities and*
> *was, I think, a gentile – though I am not sure – so that probably Bethe*
> *left Germany because of his wife's being half-Jewish. It was the*
> *multiplication of such individual cases that fortunately, very*

fortunately, lost Hitler the war. Bethe was very interesting about the
Geneva conference on atomic energy.
[Dixon in Cambridge to John in RAF on *National Service*,
18 September 1955]

That wife's father was Paul Ewald, a world figure in diffraction
of X-ray beams by crystals who, thanks to Hitler's policies, had
been brought to Cambridge by Peter and Nora Wooster (her
brother was Archer Martin, Nobel 1952). After Cambridge, Paul
moved to wartime Belfast where we knew him well. Robert
asked why, falling off his bicycle as Robert did all the time
without serious injury, he had broken his arm. Like Henry
Barcroft, he was keen to educate. 'Children fall from a lower
height so their momentum is less when they hit the ground'.
As Robert later realised, Paul had failed to understand that
increased bony brittleness with age is more important; a life-long
remembered example of how even the most distinguished
teachers overweight the insights of their specialty. Paul's mother,
Clara, was a painter. We have lost her portrait of Amélie, but
Peggy's hangs in one of our houses. Her painting of Rupert
Brooke is in the National Portrait Gallery (it is not on public
display; like Boyds in science, her painting is not perceived to be
of quite the first rank). JMcC wrote to Amélie [22 February 1944]
that Paul's wife 'Mrs Ewald jr. had to go to a nursing home for
another minor operation … but is recovering. Mrs Schrödinger
came up from Dublin to cook and look after the family. She is a
kindly woman'. This temporary cook was the Jewish wife of
Erwin Schrödinger (of the Cat and of a complicated domestic life
– Nobel 1933) who was now in Dublin, thanks to invitation from
fellow mathematician and taoiseach, de Valera. Peggy mentions

Lennoxvale being used one day for what I was told was a secret
meeting, and only my father, myself and Sarah the maid knew about
it. Paul Ewald was meeting with Schrödinger the theoretical physicist
and I believe, Niels Bohr the nuclear physicist. I am not certain as to
what they discussed, but I believe it was about keeping nuclear
findings from the Germans.
[Fink, MG *From Belfast to Belsen and Beyond*, Fink 2008]

Sadly, we have not found any direct evidence to support Bohr's participation though, as he arrived in Scotland from Denmark, via Sweden, in October 1943 and left from there for the United States a month later, he could have easily crossed briefly to Belfast during that month. Bohr knew Ewald's son-in-law, Hans Bethe, very well and, once in the USA, joined him at atomic bomb site, Los Alamos.

Friendship with the distinguished extended beyond laboratory sciences. Dixon, on a 'Committee reviewing the University of the West Indies' (we will come to it later), informs Amélie [from Mona, Jamaica, 11 March 1954] that 'when we arrived at the hotel on Tuesday a delegation was awaiting me – Ursula and John Hicks [Nobel, Economics 1972]; They are here advising the Jamaican government on taxation and had seen in the Times that I was one of the delegates on the visitation. They are very friendly, asking most kindly for you'.

Dixon, and Amélie, did not merely know such stars but were accepted by them as serious members of the wider university community, a community which extended beyond the directly academic. Dixon became chairman of the Cambridge Humanists, when it was founded (perhaps re-founded) in 1955 with a flyer explaining its aims were to foster 'discussion, to hear talks by visiting speakers, and, in general, to give expression to the humanist point of view in the City and University of Cambridge'. The 26 supporters listed in this manifesto, a veritable forest of cross-networks, include three Clare College fellows and Charles Shute as well as Fred Hoyle, EM Forster, Bertrand Russell and, note, Gwen Raverat. Dixon told us its origin was in response to the announcement of a revivalist Billy Graham mission to the university.

Less strident Cambridge Christians *were* acceptable, at least to Dixon. William Telfer, professor of divinity, thanks him for advice in his volume on *Cyril of Jerusalem and Nemesius of Emasa*! (He is not the only author to thank Dixon. So does Henry Barcroft's father, Joseph, in the preface to his *Researches on Prenatal Life* which, published in his 75th and last year, opened a new field of study. So does Neville Willmer in his *Retinal Structure and Colour Vision*).

ii   Darwins and Keyneses

Those we have just mentioned also knew each other, sometimes knew each other well. This was even more true of those related through being members of *the* quintessential Cambridge extended family. Members of it had different surnames – Darwin, Keynes, Adrian, Hill, Raverat – but were all related by marriage or by blood (more accurately Crick's DNA). Gwen Raverat was not close to us but several others of different generations were. Ideas great and small were fostered in their homes and spread across and between generations. In our time, dockers' sons often became dockers, doctors' sons become doctors, and nurses' daughters became nurses. So did academic children become academics, especially members of this dynasty.

We observed, perhaps at a meal time or in general chat, examples of how subsequent generations might be drawn into their culture of how to address intellectual or practical challenges, small or great. Three diverse examples: how to make a "back of the envelope" calculation on whether fart gas concentration would be a problem in the air of a nuclear submarine? – 'clarify your assumptions' (lunch at the Keyneses); what gives a positive academic impression? – 'make sure your answer includes an integral sign; and keep clear of smooth muscle' (Lord Adrian to his son, Richard, later FRS and a vice-chancellor); what might work methodologically? – 'roll an orange under your foot until it is squishy, cut a hole at the top and insert a sugar lump. Suck at the hole' (Margaret Hill to Robert on a bench in her garden). These are trivial, but the intellectual alertness involved was applied by many of them to bigger questions.

Ideas, and the swirl of talk, and inclusion of the young, were features apparent in many Cambridge families but perhaps especially amongst those who belonged to this academic aristocracy.

Amélie's friendship with Margaret Hill, whose husband AV (Archibald Vivian but always AV) was another prize-winner (Nobel 1922; awarded '23), followed the post-doodlebug stay at 16 Bishopswood Road. That house now bears a blue plaque for AV. She had been born a Keynes.

Amélie and Margaret corresponded regularly. Fifty letters from Margaret survive, usually mundane – 'lovely to see you on Friday, darling A, your M' – they write of children, of families, of trials and tribulations of society, of gardens, of holidays. Pretty frequently, they managed to get away from their husbands and children for outings and weekends, maybe to enjoy browsing the junk shops of Brighton and walking along its seafront, or to go to the theatre or an art exhibition. Margaret was an enthusiastic painter in oils and took Amélie along to classes. Their friendship provided an escape for both from their busy lives. They were close. Skidelsky, Maynard Keynes's three-volume biographer, describes Margaret as having a 'bisexual inclination'. It was more than inclination. Before AV, she had had a passionate affair with Eglantyne Jebb, with talk of kisses and of a shared bed and even of a marriage between them. There is no suggestion, in our letters, of physical attraction between Margaret and Amélie. Our mother had several older female friends who were, like Margaret, very close but, in those times, it was inconceivable for us to even conceive that more than friendship might be involved (we still think it unlikely!).

Margaret, though she lived in Highgate, was Cambridge through and through. She had two brothers. Maynard and Geoffrey. Geoffrey, by his own account the dunce of the family, was a surgeon who had, during the First World War, helped to introduce blood transfusion and who, long before his peers, transformed the operation for breast cancer to its modern form. Dying Frank Ramsey had, as a last resort, been transferred to his care.

When Margaret drove us to St Chad's to take up our new lives (it was by now the April after Dixon's appointment), the trio's mother was still alive. 'Old Mrs Keynes', at the age of 92 and a widow, lived on her own. Her husband, had, like Harold Taylor, been the university's senior administrator. Before that, an academic innovator, he had established the Moral Sciences Tripos. She was herself from an interesting family. Her father had been minister of Bunyan's Bedford Meeting House and her brother was an early exponent of clinical endocrinology

13. Daughters of Cambridge Dynasties – *Margaret Hill née Keynes and Gwen Raverat née Darwin. Gwen's sister was married to Margaret's brother.*

(young Amélie, while working in London, had attended his lectures). As an elected member of Cambridge City Council she served as mayor, a rare crossing of both gender and town-gown divides. Margaret followed her mother's footsteps. She and Eglantyne had, at that mother's suggestion, established a boys "Employment Registry" followed by one for girls. It too has a blue plaque. Eglantyne went on to found Save the Children. Margaret focussed on the old. Hill Homes were an early North London provision in geriatrics. Amélie, and then Biddy Barcroft, were medical advisers to the Homes.

Old Mrs Keynes

It is two days before Christmas, a grey damp afternoon in 1954, when 48-year-old Mrs Amélie Boyd drives up Harvey Road Cambridge looking for number six. In the rear of the grey Hillman Minx sit Stephen and Richard, no seat belts then. School holidays mean that they had been dragooned as non-speaking supporting cast while Mummy, very willingly, carries out Margaret's request to 'keep an eye' on her mother. A book by Defoe and a box of chocolates accompany the visit.

On the other side of Harvey Road is Fenner's. And at the end of Harvey Road rises the tall spire of the Catholic church, *Our Lady and the English Martyrs*. The old lady had been born not only before the building of that church (endowed, said HOM, by the widow of a glass bead magnate - 'dolls' eyes for idols'), but also before Fenner's, the playing field later haloed by cricketing correspondents of the *Manchester Guardian*, and the *Daily Telegraph and Morning Post*.

The occupant, when the house was located, was very friendly, warm, inquisitive, grateful. And rather impressively "put-together". 'You really will enjoy going to the Christmas Eve carol service at Kings tomorrow; it's always so beautiful to have the early evening's wintery sun glinting in as they throw open that great west door of the chapel to the echoing sound of the triumphant organ finale'. None of the three of us were religious, but, for Mummy, it was this sort of event that she had come back to Cambridge for. Such experiences were important. For us, son Maynard was also important. Not because of Bretton Woods, the IMF and the World Bank (we didn't know about them), but because of his arts theatre where we ate many excellent ice creams and later saw the first performance of *Beyond the Fringe*. 'Gentlemen, lift the seat!' - injunction above the lavatory bowl on railway trains - was irresistible when Jonathan Miller mused on its use as a toast at a gentlemen's club (or was it Peter Cook? It wasn't little Dudley Moore because, of course, he was stuck in front of the piano, and it was clearly too rude to have been the clean-cut Alan Bennett). As part of his medical course, Jonathan was, the year of the show, reading Part II Anatomy but Daddy said he had cut him some slack.

iii The Adrians

ED Adrian, our landlord, usually known by his surname, was a lord; like AV in having a Nobel, unlike in having a title (AV had refused his). He was rather small in stature for a Grand Old Man. His daughter's husband, father of Richard's Adrian, was, as already mentioned, Geoffrey Keynes' son, also named Richard, another distinguished physiologist only unusual in this circle in not being awarded the Prize. Some felt he should have been.

Hester, Lady Adrian, ED's wife, seemed relaxed about us living in her home. Had the Boyds 'by any chance located a wooden leg?'. There was no urgency, it was only the spare. Might it have been mislaid, perhaps in an attic? It was looked for, and there it was. She was grateful, arriving on her bicycle with an empty front basket and returning, rather deftly, thought Richard, with the basket now carrying the spare back to the lodge at Trinity. She was, like Margaret and Eglantyne, quite something. She had started a unit for adolescents with severe psychiatric illness at Fulbourn, the local Cambridge asylum. The unit became a focus for the development of a then-Cinderella subject. Her name did not grace a blue plaque but did have functional recognition, the *Hester Adrian Centre* established in Manchester to study the education of "mentally handicapped" children.

Both of her brothers had died in the First World War. One, close friend and possibly lover of a visiting Austrian student, had shared with him both a college room and a passion for mountaineering. Following that brother's tragic death, the Austrian student, who had fought for the other side, dedicated the book he had been writing to this lost English friend. Thus, Hester's brother became dedicatee of Wittgenstein's *Tractatus Logico-philosophicus*. The translator? Biddy's teenage brother who died on Geoffrey's ward. Cambridge networks were sometimes at a high level.

Richard, aged 11, went on holiday with a schoolfriend whose godmother, an aunt, was a medical missionary in the Sudan. As part of her very brief annual leave, she dutifully took Richard with the godson to walk in the Lake District, staying, rather implausibly, in rooms at the little pub at Seatoller. As

Richard went up its stairs, a framed, rather crumbling, newspaper-cutting caught his eye. It recounted the role played by the pub in rescue at night of a woman who, slipping in the rain, had had a horrible fall on Great Gable. The husband of the lady (she had suffered a compound fracture of her tibia) had complimented, it was reported, both the pub and the ambulance service for their assistance in getting the patient to hospital. The article mentioned his name, Dr Adrian from Cambridge. Thus, the wooden leg.

Adrian Keynes

My Adrian, Adrian Keynes (twice-grandson of the dynasty) died aged 27, a very long time ago, but I, Richard, still miss his quiet, determined demeanour and his complete loyalty. I was nine when Adrian asked me if I would like to go with him and his hamsters to spend a weekend with his grandparents. It was a large and lovely house, maybe 12 miles from Cambridge, past Fleam Dyke (a favourite Amélie-site for picnics and pasqueflowers). I was given a large bed in a spare room with, above my head and on either side, pictures of odd shapes and colours painted, I was informed, by a William Blake. Geoffrey, the grandfather, was courteous but somewhat aloof. The grandmother was different. She was small and warmly friendly and the three of us spent a weekend caring for and cosseting hamsters, eating at frequent intervals her delicious cooking from the Aga. She was, I was told later by Daddy in a significant tone, a grandchild of Charles Darwin. Whether the other 23 enjoyed hamster-mollycoddling I cannot say.

Some 10 years later that grandfather wrote a charming letter to Stephen and me in response to a (rather thin) historical article we had sent him. We had noted that an overlooked English Renaissance poem had used William Harvey's circulation of the blood as a metaphor for renewal; scientific insight incorporated, rather rapidly, into literature. Gracious encouragement from this surgical super-polymath, expert on Harvey and Blake, and Rupert Brooke's literary executor, was, well, encouraging.

Anne Keynes, Adrian's mother, I loved. She was lovely, lively and a singer. April Cantelo was her singing friend. There were brief snatches of Mozart duets downstairs. April had been married to, and then abandoned by, the conductor of Cambridge University Music Society - 'the rat' - Colin Davis.

Many years later, just before her death, I went to see a very elderly Anne. Outside, under the big copper beach tree, I told her how much I had as a child loved visiting 4 Herschel Road, round the corner from St Chad's, to play with her son under that very tree. She smiled and merely said, 'Well, Richard, my family never really did do... what do we call him... ah yes, never did do God.' And behind that comment lay, for me, the poignant thought that here was a mother reflecting on her son, my friend, Adrian, who had quite unexpectedly taken his own life some 30 years before.

iv AV Hill

AV was good, in debate or discussion, at employing humour as a tactic. 'It's the best detergent of nonsense' he had pointed out in a letter to *Nature*, in 1934, rebutting the claim of Nazi physicist, Stark, that there was no anti-Jewish discrimination in German universities. Imaginative literature or phantasy were, however, not AV Hill's metier. A biographer says that, of novels, he only liked *Lorna Doone*, an adventure set near the scene of his Devon schooldays. That Amélie chose to read that story to John and Robert is surely no coincidence. 'Hamlet? I've never seen a man like him,' was his comment to an irritated Amélie after he and Margaret had been taken to the theatre in 1945, a thank-you from the younger couple for their stay in Bishopswood Road.

As recalled by Dixon to Richard, a constitutional walk that July had taken AV to Hampstead Heath. Unusually it was on a weekday. AV had known in advance (one of a handful "in the know" about atom bomb development) of the Trinity Test at Los Alamos. He was aware of the mathematics of predicted energy release (Bethe was heavily involved) but was also understood the nature of scientific uncertainty. His walk, he reassured Dixon, had not been perturbed by any freak meteorological event.

Robert, callow youth, once asked AV if he had enjoyed being Cambridge University's war-time MP. 'Enjoy is not the right word, Robert, but I was proud to serve my country in those stirring times'. He explained the pride followed from his having achieved release of many of the refugee-scientists interned in the Isle of Man. Lightness of touch was not his style. Nevertheless, Amélie very much enjoyed her interface with his family. She reported to John, with typical zest,

> I am down with the Hills at their seaside ancient rambley farmhouse, cows just 'singing' their moos outside. You could have 'em all in the kitchen in a few moments! I came on Monday, home today (three days) and feel very refreshed. Took myself in swimming at 8 am – only me and the wader birds on their wee wheels and the oystercatchers. Y'day I was planted on Horsey Mere in a canoe by AV with Lady Geoffrey Keynes [custodian of the hamster], nice old pet, in t'other. We paddled up the Mere – it's 'like Africa' without hippos and she regaled me with stories of Sassoons and Cholmeleys. Her great moment came when AV failed to recognize a heron! ... Life is like Bergman film just occasionally; viz to find myself in a canoe with her on Horsey mere! AV is so nice a man, [illegible word] over with every mad behaviour and obsession – I also enjoyed the younger generation ... Just to add to the Bergman illusion a caravan by the bungaloid beach contains a lion guarded by a long-haired Kenya-type, 'having a rest from film work'!! I drive home via Wymondham angel-roof church this p.m.
> [Amélie at Hills' house at Sea Palling Norfolk to John, 25 June 1961]

AV excelled in the application of classical, pre-Einsteinian physics and mathematics, to physiological problems, collaborating with many scientists across physiology and across the world. One collaboration was with Henry Barcroft's father in developing an equation to describe how oxygen combines with haemoglobin. At that time, they had worked in the same Cambridge building. A more distant collaboration was with Japanese physiologist, Kazuo Furusawa. St Chad's was not the only Boyd house in which the Hills may have played a part to our benefit. Thanks to an introduction via AV, Furosawa's son became, a generation later, Stephen's long-term landlord in Osaka.

7 Anatomy and Academia

i   Research

Dixon in the lab bent over his microscope, spectacles pushed up on forehead, was, for us, "anatomy". So was his daytime absence from home and from the mainstream of family life. Organising the home, in our time, was left to Amélie. Earlier, though, anatomy had been a shared endeavour. Both got firsts at the end of their student year as BSc anatomists. Once a doctor, her limited time off from the South London Hospital was spent in libraries summarizing papers Dixon thought promising: 'Sorry about the snout reference. That isn't done yet but will be soon ... I like an excuse to go into BM [British Museum Library] with a PURPOSE'.

When his lecture duties, as a very junior Belfast anatomist, depress, she tries raise the tone.

> *Oh Dixon. I am glad you have done, are doing and going to do anatomy; what do the lectures matter at present, surely not such a great deal. After all, [lectures by demonstrators] and the general course of anatomy lectures such as Tommy gives have quite opposite aims; the former to concentrate the facts into as little time, into as certain and unchangeable form as possible, and the latter to interest you to learn anatomy ... Tommy's lectures were lovely and exciting and memorable I suppose because of T's attitude ... Later Dixon, when you can lecture from yourself – Prof Boyd! I'm certain they won't be dull. You are lucky to be doing anatomy (I do hope you feel that way still!) and I am lucky to have the chance of clinging onto the anatomical tail, somewhere right at the very end, through you. Apropos of tails: don't marsupials have well developed lips for suckling at birth, and have they a special labial nerve supply in conjunction with this? I'd love to read the beginning of your thesis, but perhaps you'd rather hold on to it yourself and bring it when you come, but if you send it to me, I promise I will treat it with great respect.*
> [Amélie, South London Hospital to Dixon at Greenisland, 22 November 1931]

Ambition for Dixon's success and confidence in its achievement are clear from early on. Lady Macbeth would have approved of one missive to him, aged 24, on his first visit to Cambridge. She writes care of Prof JT Wilson the university's then professor of anatomy [4 August 1932], 'It's nice to think of you there – you do fit you know ... How old is J.T.W? Probably been too pleasant to you to wish him dead!'.

Academics of the time were expected to teach, to contribute to scholarship, and to perform their share of institutional administration. Only some engaged in research. In his chosen discipline, Dixon did all four, creating, claimed William Hamilton in an obituary, 'an atmosphere in his department which led to research being regarded as an absorbing aim'. Certainly, Dixon was absorbed.

Anatomical research, the nature of biological structures rather than of their function, was unfashionable when Robert and Richard became medical students in the '50s and '60s; less so when, in 1930, Dixon, inspired by Walmsley, made his choice. But even then, it was seen by many to be less attractive than the rival discipline of physiology with its focus on how things worked. Study of anatomy, the core of medical learning since Vesalius, had, to some extent, split during the 19th century. One path led to the role of structure in understanding biology, the world of Darwin, the other maintained anatomy as handmaid of clinical, especially surgical, practice. His research into two topics established the young Dixon as a rising star. One, 'Classification of the Lip', of which more in a moment, followed the former path of intellectual biology; the other, 'The Development of the Carotid Body', was a little closer to the clinical. The carotid body had recently been found to be critical in the control of breathing; the adjacent carotid sinus similarly important for blood pressure. His year in Baltimore was spent on the carotid body. Reading one of his early articles, we are surprised to learn that he had, by then, also done a minor piece of direct clinical research involving the sinus. Pressing it, he found, lowered blood pressure less in those with, than those without, an abnormally high pressure – hypertension. At that time severe, "malignant" hypertension was

174

a fatal disorder. His subsequent demonstration, in collaboration with a neurosurgeon and a junior colleague, that a minor anatomical feature of the autonomic nervous system would undermine the effectiveness of surgery on autonomic nerves as a treatment for hypertension was in the same tradition. The findings might have become important for patients had not drugs capable of lowering the pressure then supplanted surgery. Perhaps he had more of a hankering to improve clinical practice than we realised.

Overwhelmingly though, the topics, then and later, which drew his scientific gaze related to understanding rather than application. His studies almost always also had a developmental slant; how the "embryology" of an organ can explain its structure and contribute to an understanding of how it functions. Gaze is the right word, his approach depended on looking down a microscope. 'Dixon Boyd (wrote another obituarist), had an extraordinary ability to visualise and interpret the intricacies of

14. Dixon and microscope – *polishing his glasses, a typical act; about 1960.*

anatomy as revealed under the microscope ... it was one of his most memorable scientific attributes'.

Microscopic techniques were central to his first publication, 'The classification of the upper lip in mammals'. It was accepted by the *Journal of Anatomy.* Walmsley is thanked in it. So is Le Gros Clarke (which was misspelt, it should have been Clark), the journal's editor, who was to play an important part in Dixon's professional life. The paper is a tough read, full of technical specialised vocabulary of the time. Yet, the simple fact that lips are central to feeding renders their classification relevant. Dixon's finding of five conceptually distinct ways of grouping mammals when considering both how the top lip is arranged and how it develops was both novel and completely compatible with what Darwin might have recognised as a pebble in the edifice of evidence supporting mammalian evolution. The paper also sheds light on how a human abnormality, "hare lip", arises during an individual's development.

Amélie shared delight that Dixon's presentation of the work to the Anatomical Society was 'a success. I am so glad – for us, for I do appreciate how much hangs on the laps of [the] Elliott Smiths and the J.P. Hills'. Approval of one's early work by such disciplinary grandees was, as now, professionally important. Despite growing distractions – the road to war; bombing of London; conscription of colleagues; evacuation of department; doodlebug damage to lab and home; separation from wife then arrival of new babies – ideas seem to have cascaded from the fertile mind of the young researcher, then London University professor. He published on a wide range of research topics, sometimes with younger colleagues. He also found time to became co-author of *Hamilton, Boyd and Mossman.* It was translated into several languages and gave his name to generations of medical students.

Later, his personal research was increasingly devoted to one issue, elucidating the anatomy of the human placenta ('placenta in Man'[sic]). The placenta was, as we have seen, an interest shared by William and most of the papers which resulted and their monograph, 'The Human Placenta', bore both their names. Research on other topics in his Cambridge department was fostered through the support and encouragement of those he recruited (and, more challengingly, those left over from the regime of his predecessor). Thanks in their papers provides the evidence.

Placentas could be obtained after birth. They are, after all, the after-birth. More useful for research were placentas obtained still attached to a surrounding uterus. They were obtained following emergency hysterectomy to staunch uncontrolled bleeding or *post mortem* after death of a pregnant woman. Arrival at the department of attached '*in situ* specimens' allowed the unborn child and his or her (in those days, *its*) placenta to be studied with connections to the mother intact. Dixon eventually accumulated a very substantial number of such specimens now known, but not then, as *'The Boyd Collection'*.

In Dixon's publications, obstetricians (including Alice Townsley) are thanked for their collaboration. There is no mention of the women involved, let alone of consent from them or their widowers.

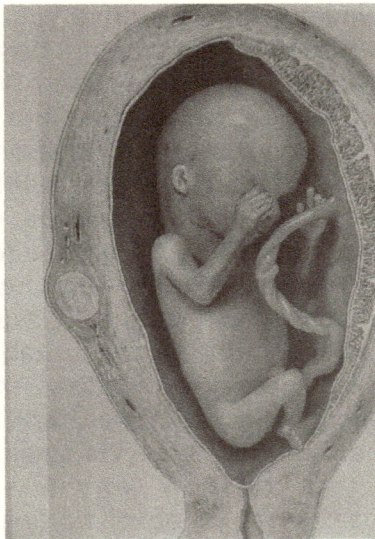

15. In the womb – *in situ embryo attached to placenta (on right) then uterus; from the 'Boyd collection' 2024*

Goethe wrote that 'method is everything', a favourite German phrase of Dixon. Nevertheless, in 1932 he had written to Amélie, that to 'be content with scientific technology is the most terrible thing'. His preferred position was to be an ideas-researcher who used tested approaches. He did not wish to be driven by every new tool which might become available. In the post-war years another generation was to apply new technologies – functional, immunological, chemical. Although he arranged for electron-microscopy to be introduced in the department and supported development of histochemistry and tissue culture, Dixon did not fully harness the methodological revolution. Perhaps that was a mistake.

Richard as chauffeur

I had just passed my driving test, so perhaps 1962. 'Amélie, I need to be at the BBC Cambridge studio this evening, Medawar doesn't want to do the interview, and has suggested my name instead.'

'Why don't you get Ricky to drive you down.'

He and I proceeded, ending up in a tiny room upstairs next to the Amateur Dramatic Club (the ADC theatre).

Very casual, a red light went on. 'And now Professor Boyd, thank you so very much for making time and coming to help listeners understand the possibilities opened up by this new work done in London; do you think the research has the potential to change life for babies born prematurely?' No wonder Medawar had the nous to avoid something with 'clinical implications'. Daddy, I thought, did a good job in responding to the suitably sycophantic middle-aged male interviewer. How important and interesting human development was, and that research progress in the field tended to be via small increments in knowledge, and that this new work was significant as exactly such a useful set of novel findings. As we left, interview over, the interviewer said, 'There will of course be a small cheque sent out in due course, Professor, and how very nice to meet you and your son.' Daddy had a sherry when we got home.

Medawar had been recently awarded the Nobel Prize and I had written an essay for school on his work - 'What is the immune system?' That Medawar himself had recommended the BBC to interview Daddy certainly impressed impressionable me.

Research did not swamp Dixon's other academic roles. The monograph and textbook then competed on more equal terms with the research paper, and so did teaching. Funding was not research-dependent and neither did research depend on successful grant application. We have no evidence that Dixon applied for any personal research grant after his Rockefeller Scholarship. According to its archive, and astonishing to us, the Medical Research Council only made one allocation to any group in Cambridge in his entire time (to the Strangeways Laboratory; application supported by Boyd). Overwhelmingly, the university's "departmental allocation" covered the costs of staff and their research. Some other sources might be tapped into – Joseph Barcroft had funds from the Agricultural Research Council, and the Rockefeller – but anatomy, so far as we can establish, received no external funding apart from one Welcome Trust grant.

Dixon died in 1968. A Cambridge biologist wrote 40 years later that

> when Harrison [Dixon's successor at the London; then in Cambridge] was appointed, the Department of Anatomy unlike the Department of Physiology had a reputation in research that was far from being internationally acclaimed. Harrison began a process of transition and the department is now a leading performer on the international stage. Harrison had foresight in appointing young, very able research scientists who had very little knowledge of topographical anatomy but were willing to learn. He also encouraged the careers of those who began life as technicians rather than as academics and established research positions for them in the department.
> [Kaverne E. B., Harrison R.J., Royal Society Obituary, 2005]

We are doubtless biased but do note that reputations can rise and fall. Within two decades, the renown of Cambridge Physiology, which Kaverne extolls, slumped dramatically in the nation's first semi-objective (very semi-) research assessment. Also, a scientific backwater can become important. Long after Dixon's time, there was a resurgence of interest in the developing structure of embryos which now, as "developmental biology" became a major field of medical research. As part of this refocus, studies on molecular and cellular developments in the placenta began to be published. Quite a number have acknowledged use of material from the Boyd Collection.

Walter Bodmer — the 'New Science'

It was a big, brand-new Xerox machine. But it required the user to remember to remove what had been copied. Uncollected in the in-tray were ten pages - from Boyd and Hamilton's *The Human Placenta* - with Walter Bodmer's name adjacent. Why was Walter Bodmer, the 35-year-young new professor of genetics fresh from Stamford, interested in Daddy's obscure organ? Why was the human embryo (carrying genes from father as well as mother) not rejected as "foreign" by the mother's immune system? Bodmer, the ultra-modern, somewhat charismatic, experimentalist, was studying what Dixon and William - two boys from Walmsley's BSc class - had to say on the question. Delivery of the pages was gratefully received.

But, others of our father's time appear to have shared Kaverne's view. Unusually for head of a bioscience department, Dixon was not elected FRS.

ii Professorial Politics

Le Gros Clark in Oxford, Zuckerman in Birmingham and our father in Cambridge came to dominate English anatomy. All three are medical doctors, research workers, teachers, examiners and writers of successful text books. Dixon's handwriting is the most illegible, Le Gros has the smallest neatest script, Solly displays the boldest flowing hand. Dixon was friends with the other two.

> *30th July 1945*
> *My dear Dixon, Would you have a glance at this paper of mine ...*
> *Best wishes, W.LeG. C.*
> [Wilfred le Gros Clark]

> *20th June 1949*
> *My dear Dixon,*
> *I was very moved by your note ... You have been of enormous help to me in launching our experiment [new anatomy course] – and in launching me [anatomy chair in Birmingham 1946-68]. Your encouragement has meant far more than I've tried to express ... let us continue to see each other as often as we can. Joan sends her love. Yours Solly [Zuckerman].*

Relations between Solly and Le Gros deteriorated. The rift nominally arose from a philosophical divergence about scientific method; how the shape of pre-human teeth (key to understanding the origin of *Homo sapiens*) should be interpreted. In truth, their mutual dislike is also about status. Solly had, in 1934, been recruited, young academic, into Le Gros' department in Oxford where he worked with, and published with, his boss. With the war he added to his work as anatomist a contribution to the war effort – "operations research". (AV had pioneered such research in the previous war). This led to the quite extraordinary rise of his becoming a national, even international, political sage and influencer of governments. His marriage to Joan, the daughter of a Marquis, may not have been a disadvantage.

The dental fight between Le Gros and Solly led each, in the early 1950s, to recruit heavyweight statisticians: Jacob Bronowski FRS from the Coal Board for Le Gros; Frank Yates FRS from Rothamsted for Solly. Perhaps because of an accusation that Zuckerman's statistical "estimates of variance" were in error, the hostility between the protagonists became bitter. It continued post mortem in Solly's long and sometimes vicious Royal Society obituary of Le Gros – the pursuit of academic conflict by other means: 'Had he not been so sensitive to comment which was adverse to his own views, the reverberations of the posthumous and accidental demolition of the Australopithecine thesis which he maintained and expounded so forcibly right up to the time of his death, might prove to be far less reaching than they are likely to become ... He wanted to be admired, and it rankled when some whose esteem he sought ... denied it'. This antagonism from one who, continuing as anatomist, became chief scientific adviser first to the Ministry of Defence and then to the government (becoming a key figure in development of the 'Treaty Banning Nuclear Weapon Tests in the Atmosphere, in Outer Space and Under Water', of the Polaris Agreement and, more parochially, of plans to build the Thames Barrage) is odd. For Dixon, a nice note from Solly in 1953 asking for a list of his recent publications so as to propose him for fellowship of the Royal Society did not lead to success. Did reluctance by Le Gros to partner with Solly in supporting the candidature contribute to its failure? However – a favourite word of Dixon's – his style of research may merely not have been seen as sufficiently innovative to meet the expectations of an FRS. Failure to be elected became an ongoing disappointment ameliorated, perhaps, by his expressed belief that some colleagues felt similar failure at not being awarded a Nobel Prize.

Richard rejects a suggestion Dixon made to us that the antisemitism of elderly professor of pathology, "Daddy" Dean, had contributed to him rather than Solly being appointed to the Cambridge chair. (Pre-dating the introduction of a retirement age, Professor Dean continued to occupy the pathology chair into his 80s – 'I know I'm a scandal, I intend to go on until I am an abomination' – and found time to coach Robert for a scholarship post-Cambridge giving rather good advice on how to draw information from a pathology-slide). Richard's view is that, although anti-Jewish sentiment was not absent from academic politics of the time, Zuckerman's wobbly reputation as an experimental scientist, FRS or not, scientific adviser or not, was a more important reason.

If not Cambridge, might Solly get the Oxford chair? Le Gros was to retire so Dixon travelled to Oxford to talk privately with the "regius" [professor of medicine, more correctly *physic*] about possible candidates. A letter to John describing that visit to Oxford does not, in fact, mention its purpose but is worth quoting for the flavour it gives of Dixon's attractive laterality of thought (and relation with sons).

Wednesday, 27th *June, 61*
My dear John,
... Can you spot the author of this: "the thing about Proust is his combination of the utmost sensibility with the utmost tenacity. He searches out these butterfly shades to the last grain. He is as tough as catgut and as evanescent as a butterflies bloom." Don't you agree she has put her own sensitive finger right on the spot? It is, of course, Virginia Woolf. I was in Oxford yesterday and had to wait for a while in the room of the Regius Professor of Physic; to put in time I took a book from the shelves. V.W.'s diary it was and, as a sortes Virgiliania, as it were, I opened it at this comment and, of course, at once thought of ... your mention of how you associated what you had read with where you had read it. I, too, have had a most strong evocation of the past when re-reading something which has moved me and which moves me again. I daresay it is a not-uncommon experience. But that's not my present point. As I was putting the quoted comment into my pocket diary I was surprised (old sense) by Pickering, the Regius, who said "oh, Virginia Woolf. That's strange you should have chosen that volume from so many. I hadn't thought of her for a very long time and two nights ago I had the most unlikely dream imaginable. I dreamt I was in a library (actually my sister-in-law's) and there was a shelf of books which had obviously come from Virginia Woolf's own library and on the cover of each of them in bold golden letters was her name. And in vulgarity, of course lies the unlikelihood for can you think of anyone less liable to treat her library that way than Virginia Woolf? But why should I have had that conceit in my dream?" Of course I could not help, but when I got home and told your Ma she at once, with her feminine common sense, which goes with me for clairvoyance, remarked that if we (i.e. she) knew the sister-in-law, the answer would probably be forthcoming at once. But as for you, poor John, you mention Proust and you get thrown back at you a Regius professor's dream but Proust would have liked the whole little anecdote for each one in it, save perhaps the sister-in-law, shows some sensibility though I may have been over-tenacious in recounting it.

The subsequent electoral board, with Boyd as "external" went for the GW Harris who had earlier hoped for Dixon's chair in Cambridge. Boyd found him a better (and perhaps more reliable) experimentalist than Zuckerman. The subsequent development from Oxford anatomy of, arguably, the entire field of neuroendocrinology confirms the wisdom of the choice and makes one wonder if Cambridge should have appointed that other Harris, rather than Boyd, to succeed HA Harris ten years earlier.

iii Teaching

The family found Dixon a good teacher. Amélie, as fellow student, had valued his explanations and for us, at least for the young John and Robert, an evening walk could include, 'Can we do general knowledge Daddy?' which opened a mutually enjoyable exchange. 'How does light come from that street lamp?' was a typical question. Did it come as a straight line, or in waves, and how fast, and how long would it take to arrive from the lamp's apparently adjacent moon? His style was enthusiastic and clear, with facts placed in a wider context. Letters suggest his teaching was also appreciated by others.

> *Dear Professor Boyd ... I am sure the others will agree with me that all credit is due to your magnificent teaching and the efficiency of the anatomy department. I always think that teaching a crowd of students must be a thankless task, for no sooner do they obtain an intelligent grasp of the subject than they go, and a fresh lot arrive. However, I hope you will obtain a little satisfaction from the fact that you were able to raise four, not over-intelligent students, to concert pitch in under two years in Cambridge, in a war.*
> [Bruce Bomford, surgeon-to-be, to Dixon, 15 June 1940]

This was two weeks after Dunkirk.

Despite his initial fear of the lecture theatre, teaching could be a delight. 'I lecture to-morrow [May 1941] on the medulla about which I know quite a lot so it will be fun'. But not always, the next month he is oppressed by 'the futility of doing anatomy in these times, a feeling of complete lack of physical energy and weather which is just like December. God damn the weather'.

Echoes of good teaching can resonate years later. Reminiscences, in a copy of the London Hospital alumni magazine dated a half-century after he had left for Cambridge, recall him as 'the teacher I best remember'. Widowed Amélie, while staying with John, met a leprosy surgeon at a small hospital in Hong Kong's New Territories whose daily operating was the reconstruction of hands. He had, he said, only gone into hand surgery because he was failed in a *viva* on the hand 'by Prof Boyd'. He had been told that, to pass, he must understand, really understand, how the hand worked – 'innervation, the actions of the long and short flexors, and of the extensors, the abductors, the adductors' – the fruits of that initial failure underpinned, he said, his daily work. Pearce, a doyen of histochemistry, giving a plenary lecture to the Physiological Society, pronounced that all APUD cells (a then hot category; his concept) were of neural crest origin, 'and I learned about neural crest from my Professor when I was a Cambridge preclinical student. Boyd was an inspirational teacher'. This statement was to Richard's utter astonishment, and pleasure. Pearce certainly had no idea that the youngster sitting in the back row was a Boyd son.

Dixon, in mid-century Cambridge, was not unusual in having substantial teaching duties. The professors of physiology and pharmacology also taught elementary courses and, in Belfast in the 1930s, Henry Barcroft covered the entire first year physiology course in a series of daily nine o'clock lectures delivered by himself. He now has a QUB blue plaque, though maybe not for that. By contrast, as a student in the late '50s, Robert does not remember the Cambridge professor of biochemistry lecturing. Academic culture in that new discipline was perhaps already moving towards the expectation that, in status and preferment, research should trump teaching. Dixon, professor in medicine's oldest discipline, did not see himself as just a researcher.

iv Scholarship

"Keeping up with the literature" was a daily commitment involving monographs and *reprints* of papers solicited from their authors; no photocopies then. *Nature*, and other general scientific and medical publications were a weekly read underpinned by regular perusal of disciplinary journals in anatomy. For a while, Dixon edited the *Journal of Anatomy*.

There were meetings.

I am very preoccupied – Anatomical meeting here next Thursday, Friday and Saturday, meeting in London on electron-microscopy on Monday next and the following week I am to open a symposium in Oxford – a very high-powered affair. I am a little frightened – but not for the first time!
[Dixon to John in Göttingen, learning German, 28 June 1954]

The Oxford meeting was to try to achieve consensus on the structure of "spindles" in muscle; adjudication being another role of senior academics. There were invited lectures, and colloquia and conferences, small and large, though fewer and less sub-specialised than in subsequent decades. 'I have just received an Invitation [30 September 1955], deserving the capital [letter], to go to a conference in the States next year – at Princeton in March. The sugar is a grant $1,000 (!) to cover expenses. This academic racket sometimes has its pickings'. Travelling across the Atlantic was getting quicker but remained complicated. 'I return Boston to Halifax, Nova Scotia (where I lecture), Halifax to Gander, Gander to Shannon, Shannon to London'.

Our period covers a time when colonies were becoming independent countries. Dixon's letters and tax returns indicate that academic support and mentoring across those colonies was an ongoing expectation of Cambridge professors. So was advising on their academic appointments. Several of those who worked with Dixon continued careers across the Commonwealth. Of the ones we knew, Grillo became dean of medicine at Nigeria's Ibadan University and College Principal in Sierra Leone and, briefly, fellow of Churchill College, Kanagasuntheram was later professor of anatomy in Singapore while Navaratnam remained to become a lifelong Cambridge academic. At a more junior level, about half of candidates recorded as paying fees for PhD examination by Dixon have names suggesting "New Empire" origin. Examining medical students in Ibadan and in Khartoum was, he told us, to maintain standards of medical graduation so as to allow continued recognition by the General Medical Council. Travelling overseas to examine was fun as well as 'contributing to development'.

Dixon's grandest Commonwealth involvement was as member of a heavyweight delegation considering the future of the University of the West Indies (that J.H. Harper from his London department was now its professor of anatomy may have helped). It came with 'perks'.

> *A grant from the Colonial Office for £40 for tropical kit! Imagine me in Tussore and Panama hat! William [Hamilton] characteristically got me to the plane in good time after a good breakfast. There I met Dame L. who seemed half dressed. She had overslept at hotel and come straight to airport instead of going to London terminus. The P[ost] O[ffice] had not phoned her. Then the tickets were all wrong (Colonial Office mistake). Then arrived Sir Irvine [retired VC, Sheffield University] and Lady Masson (neither of whom had flown before) and Adams of the inter-University Council [Adams had worked for AV Hill at the Academic Assistance Council and later, in the turbulent 60s, became director of the LSE].*
>
> *Yesterday we went over the whole university and hospital ... considering that the whole concept is only 10 years old, [it] is good and the hospital quite fascinating ...The grant flow, of course, is why we are here – finance, and from what the Hicks say it will not be easy to continue the development which is an infinite pity.*
> [Dixon, Mona Hotel, Kingston, Jamaica to Amélie, 16 March 1954]

Despite the Hicks's pessimism, the university's current website is upbeat. 'Starting as a University College of London in Jamaica with 33 medical students in 1948, the UWI has evolved into a modern, future-driven, activist, top-ranked academy with over 50,000 students'.

Dixon's more general comments on Jamaica are of the time.

> *We had dinner with the Governor on Friday and then yesterday. The whole situation is hot with politics, and of a particularly complicated sort – very reminiscent, in fact, in many ways of Passage to India. But*

187

while it could be funny or pathetic for a novelist it is tragic when it impedes the future of higher education and the medical services in the islands of the Caribbean.

> *The Governor, Sir Hugh Forte [sic, actually Foot], is an excellent person ... He, and his wife, seem really to have entered into the spirit of colonial reform, they know the Island and its people and they really want to help both. And the people here obviously like them. The trip yesterday was quite impressive ... We passed villages of extraordinary poverty with little bright faced piccanninies (but, as I learnt to my cost, one does not use the term here) playing cricket. They all take the Test matches so seriously and they get such satisfaction out of the prowess of the West Indians who do better than the English. But ... success is not enough to place in the scales against West Europe and its civilization.*
> [Dixon, Mona Hotel to Amélie, 21? March 1954]

Within Britain, teaching for other institutions was a useful source of income – '£90 for the nine lectures in Anatomy' as Professor to the Royal Academy (a post initiated by the great 18th century anatomist, William Hunter. The other William, William Hamilton, succeeded on Dixon's departure for Cambridge). So was examining. 'Have been appointed [1941] examiner in Anatomy by Royal College of Physicians! ... it will mean £100 for five years'. Later, there were leadership duties, and prestige, as president of the Anatomical Society. Beyond anatomy, there were administrative tasks: vice-dean of the London, Senate Council member in Cambridge.

Days which could include teaching, book-writing, administration and college life, as well as helping Scotland Yard find murderers, don't suggest he was exclusively focussed on his research. Finally landing the fish was always a challenge:

> *I must finish two papers for Le Gros before the end of the month (it is so characteristic of me to have them each half finished) and term starts next Monday.*
> [To John, 4 December 1960]

v Clare College

Clare was family for Dixon, and for Amélie, as it was for us. Initially Dixon had been a tutorial fellow with teaching duties for the college. Later he was, less onerously, a professorial fellow. Just as Nanny, Norah and Hedy wound lifelong through all our lives, so did Clare. We socialised with Clare children, the parents dined with other fellows and their wives (Amélie was outraged by Teddy [Sir Edward] Bullard – 'one of the most intelligent people I know' - laying a hand on her knee at one such dinner. She was also irritated when Daddy danced too enthusiastically with fellow Clare-wife, Penny Willmer. Family indeed!). Of the young academics participating in the ballet outing of 1936, Greaves (rather than Richard's father-in -law) became professor of pathology, Spooner was (despite his alleged antisemitism) alongside Dixon on the shortlist for the mastership when Thirks

16. Fellows, and some wives, processing over Clare Bridge –
"Thirks", Sir Henry Thirkill, in front (fourth from left),
Dixon behind (tall, seventh from left)
image © Lafayette Photography Ltd

retired, and Hammond, classicist and war hero, gave Amélie's memorial oration in Clare Chapel 62 years later.

By the time the flag by the porter's lodge hung at half-mast to mark Dixon's death, Clare's fellowship had become larger and less familial. When he was first elected there were 12 fellows, today more than ten times as many share that role.

Clare is a frequent mention in letters mostly as somewhere to entertain visitors or as a place to stay when Amélie is away. But, events after Thirks' retirement makes it plain that our father was thought to be a significant "college figure".

> The meeting at Clare went on until 12. A shortlist of six [for master] was drawn up (this is _very_ much to be kept to yourself). Cockroft, Ashby, Page, Spooner, Baker, Boyd. This was on a secret ballot. It includes, you will see, neither Harry [Godwin] nor Harold [Taylor]. I think this is very hard and I am sure they are very hurt [at least Joan Taylor was, she complained to Amélie that Dixon had betrayed Harold]. As for me I at once said I was not a candidate, as I have indicated before. So, the list is five, now. I dare say there will be some changes as time goes on. All I know is that it is _not the job for us_. I have, of course, got some pleasure out of still being considered a possibility [as he did at being 'pressed by Brian Reddaway on behalf of the younger fellows not to refuse'] but I must _not_ weaken. Neither of us would like the chores and the [masters] Lodge though beautiful at night is dark during the day. Besides I am an Anatomist and I don't think the jobs can be mixed with any fairness to each.
> [Dixon to Amélie, 9 May 1957]

Dixon probably suggested the actual appointee, Eric Ashby, Lennoxvale-neighbour of his father-in-law and QUB's vice-chancellor. Ashby became a radical master. During his tenure, women became welcome both as students and as fellows and Clare Hall was founded to become a separate institution. Neither move was popular with all the fellows though they were with Dixon – family disagreement again.

A core question, recurrent over the decades, is what, if anything a college brings to academic endeavour beyond the social. Did Clare add to Dixon's research or other activities? A college environment, if fellows participate, clearly brings diverse disciplines into conjunction and Clare's, then small, fellowship encouraged participation. Dixon took a lively interest in the work of other fellows in fields as disparate as the *Fermi Surface* – Pippard; *Early Christian Fathers* – Telfer, or *The Retina* – Willmer. More speculative is whether such High Table chat contributed to anatomy. Plausibly it sometimes did. Dixon's introduction of histochemistry involved identifying the appropriate talent to work with Charles Shute. The recruited individual, Peter Lewis, was probably suggested by Willmer. Anatomy's development of electron microscopy involved participation by Cavendish expert, Cosslett, introduced to Dixon by physicist-fellow, Pippard.

It might have been even better for his science, thinks Richard, had Clare had a chemistry fellow for Dixon to sit next to in Hall. Clare had had one but he had resigned his chair. That individual, a plant chemist, felt the Department of Biochemistry required more medically-oriented leadership. He was wrong. His protegee Sanger's techniques for "sequencing" molecules has arguably had a greater impact on 21st century medicine than that of any other scientist, not excluding Crick. That chemist was Chibnall, Chibbie to Amélie. He had been Sanger's PhD examiner and encouraged him to use a key chemical – FDNB – from which all else followed. Chibnall, it is said, declined to be an author on the crucial paper. Chemistry or not, Clare probably did make a genuine contribution to research in the anatomy department.

Dixon might have liked us to make another point not restricted to his Clare life. In contrast to Rutherford who famously threw an administrator's letter aside unopened with a dismissive 'such a nice fellow, such a nice fellow', Dixon admired effective administrators such as Thirks, or Hugh Foot, or Walter Adams, or Clark-Kennedy (his dean at the London). He was, himself, regarded as having useful administrative skills as indicated by his having been placed in charge of evacuation of his medical school as vice-dean and being later considered as a plausible master of Clare. The letters draw attention to one other leadership role Dixon might have taken up, director of the Carnegie Institute, where he had spent such a scientifically productive 1935. He was in 1955 in the USA in Boston visiting Harvard anatomist, George Wislocki and fellow placentologist, Ed Dempsey, in St Louis. Amélie writes to him there, the letter misses and is forwarded. 'Am terribly interested in your experiences and especially that you are being refreshed mentally by the Yanks. Must say [Harvard anatomist] Wislocki's goodness is nice – and a repost [*sic* – riposte?] to all scathing anti-Americanism. But it's good we voted against the Carnegie for us. Not (as Hansi [of the German tutorials; English not her first language] so nicely puts it) 'our tea cup' but the equipment is interesting and you will start next term with new notions I am sure'.

He did have new notions. Not the Carnegie, but rather a major expansion into placental structure using the electron microscope developed thanks to Coslett. He writes to teenage John, then earmarked for science, a letter with sketches of the microscopic findings.

> *Arthur Hughes and I have been doing some work on the human placenta using electron-microscopy (E.M.). The problem is mine and the technique his (with the help of the Cavendish), the observations, so far, very interesting. We are concerned chiefly with the outer membrane of the foetal part of the placenta – the semi-permeable membrane across which all things must pass from mother. At ordinary light microscopic magnification (which goes down to about one micron) this membrane has, from point to point, along its complicated surface some appearance of villus like projections [sketch figure], like a fence[?] or brush – hence called 'brush border'. With the E.M., however, we can resolve this border into a most complicated series of protoplasmic projections [figure]. Naturally this increases surface area enormously <u>but</u>, and this may be important, under those areas with no projections we find a number of vacuoles and our suggestion is that these processes may be engulfing these vacuoles from the maternal serum, thus [figures] In this way the placenta may be getting things across other than by the usually associated passage of molecules across semi permeable membranes – which would explain some discrepancies in the things that do get across. I may say that such a phenomenon was established for the passage of <u>some</u> fluid and substances across the surface membrane of amoeba or of isolated cells in tissue culture. It was first suggested by Warren Lewis who worked in the Carnegie Laboratory when I was there and who called*

*the phenomena "pinocytosis". Our findings are highly suggestive
that a similar mechanism may be operative in the placenta. Sorry to
have inflicted it on you but I am in rather full of it just now. And in
due course you will see the photographs.*
[Dixon to John, 20 February 1954]

(It is something of a pity that placental microvilli never hit the
jackpot. Crick's more or less contemporaneous and similarly-
illustrated letter to his son was sold in 2013 for $5.3M).

To learn of the Carnegie offer was a surprise to Robert,
but not to Richard to whom it must have been mentioned. At that
time, he was collaborating with Dixon in further work on
placental microstructure which they reported in several papers; a
Cambridge inter-generational tradition, Henry Barcroft as a
student had published with his father and there are other examples.
The Carnegie directorship declined, Dixon's career remained
firmly in Cambridge where placental studies continued until
death intervened 13 years later. *The Human Placenta* with
William, was to be a posthumously published goodbye.

8 Amélie and Dixon

i   A Cambridge Wife

Reading Amélie's numerous, often weekly, writings to John and Stephen – at boarding school, then living abroad – reminds us of the forcefield of her personality. One typical letter (typically long too, 2000 words and selected almost at random) examples how she expended her days and her energies; the social life of a professorial wife. Her contemporaries, wives of successful Cambridge figures – Joan Taylor, Barbara Reddaway, Anne Keynes, Helen de Frietas – didn't seem quite so manic as our mother but they all similarly participated in Cambridge life.

At the time of this particular letter, Amélie is 55; John is in California on vacation from Yale where he held a postgraduate fellowship after Clare.

4th Aug. '61   21 Newton Road Cambridge
My dearest John
Sorry for long gap ... Jill [Thistlethwaite, daughter of Frank, first VC of University of East Anglia and "girl-friend" of John Taylor whose father Harold, having failed to become master of Clare, left to become first VC, in his case of Keele University. John Boyd had a tendressse for Jill] came for lunch and seemed to add so much more ... In Norfolk, staying at Hills' house at Sea Palling. I rented a half-decker for Rob, Rick and Steve and they had 2 sunlit windy days. We got into the sea, too ... I got for you a rather dreary but true Scotch sheepy lot of socks. 8pm and they await some transport sign of life. Could I use Ann Yale [daughter of JMcC's Belfast solicitor, married to Cambridge academic about to move to Yale]? No sooner thought of: phoned – she is coming for supper! All this took half an hour. I made stew and laid table and now having organised the socks in such magic style I must try the shoes in good hope of success! The Yales leave at 6 am tomorrow so that was a close shave!

It is so good for you to see San Francisco through kind Lenways and I am v. grateful. I already wrote to Fred [her cousin, formerly Fritz Lewandowski, now Americanised Californian businessman] but wonder if I posted (RSVP). How are the little girls – if you gave me arm length and neck and waist at back I could send three nice Fairisle jerseys? ... Do try and visit the Iklés at Santa Monica [Judith's brother, (now Fred rather than Fritz) an increasingly successful sociologist of war and later Republican Under Secretary of Defence] ... The Shutes [Charles and family] came back 2 weeks ago. I was written to and asked to collect keys, groceries etc. The letter came on a Thursday with early closing and we were leaving for Norfolk on Friday early. I baked and shopped and fluffed about in my normal silly way and, so far, have had not a peep from them. But let that pass ... Rick is in Skye with rucksack and nice Leys friend. Armed with maps and mother fuss. I left him and his pal in the front of a lorry to hitch to Edinburgh ... Steve had 12 music makers for 4 days, 5 of them resident. ...

Patrick [Gowers, Clare musician and friend of John] is wedding on September 23rd and 'not through expediency', he writes. I will see him and Caroline next week, I hope, before leaving for France. I will give them 'Guide Bleu' to France, I think plus a 'do-da' ...

Robert was to see Beresford [Davies, psychiatrist re treatment for homosexuality] yesterday and decide whether he wanted to spend my mother's inheritance on a full analysis at the Tavistock. It's his decision. I wouldn't if it were mine: analysis is not doing its job say many to me, whose judgement I think valuable ...

[Rick] works hard and is much matured, you will find. Steve is still quite unable to do this. His efforts at 'working' are quite unsuccessful ... Knowing how much more ability in real sense he has than Rick I grieve to see him fritter all, à la Pa. Whatever ability one has there's got to be a cement of will – or enthusiasm. Steve is such a university person that it's 'wicked waste' to see him thus lose his exams [he didn't]. ... I was about to tear this page up but now having 'let fly' I feel quite calm again, so will let it fizzle out with you! Dear John. Wait till you have your 10 little boys: how I will enjoy them and 'laff' at you fussing over their welfare. ...

I am beginning to think of our French beach with delight ... no cowslips this time ... I shall take Bellow's Henderson the Rainmaker' [sic] ... V.v.v. much love. What a terrible letter! X Ma.

'Came for lunch...' 'cowslips...' 'No sooner thought of...' 'I shall take Bellow...' 'fritter all, à la Pa...' 'Having 'let fly' I feel quite calm again...'. Each of these words or phrases is, for us, redolent of her personality and of her life. Each remind us of examples in other letters:

'Came for lunch...'

Her days were full with hospitality and involvement in our and other young people's lives. 'Rob had the U.C.H. hockey team down for the weekend and about half of them stayed'. Our friends were drawn – sucked would not be too strong a word – into Amélie's orbit: 'Nick Forbes was here today. He has returned from two months roaming through Italy ... Bill H[umphrey, son of Leys headmaster] much perturbed and coming for daily advice'. Adults and young were often intermingled as at an '"overcrowded setting" lunch, we had Ann and David Yale, who go to Yale in September for a year as visiting law Prof ... nice Jo Vining [US Clare postgraduate] and his superlatively pretty fiancée – wicked James Bartley [J.O.B.] and altogether much jollity. J.V. [undergraduate Doggart son] rather starry eyed over Yeats (puts roses on his grave each summer) rapidly disabused by J.O.B.s bawdy details of the bard'. Some adult guests were grand or grand-to-be: 'We are having E.M. Forster here to dine ... Kingsley Amis for lunch yesterday ... diffident quick and bright'. She enjoyed their minds and personalities but was, perhaps, a little more susceptible to their status than she would have liked to admit. Even in student days, 'Maurice Ravel ... was rather fun'.

Raverat's university dinners may have gone, but Clare and departmental entertaining – Amélie as hostess at home or guest elsewhere – had not.

> *Monday: the Department (17 of them) – to a delightful flic – "Tirer le pianist" (Truffaut) with Charles Aznavour. The best since "Citizen Kane". Full of beautiful "shots" and suggestive, original interpretation. Do see. Afterwards home for supper – smoked salmon etc – not eaten by Whiskers [the cat].*
>
> *Tuesday: the dear Doggarts for Clare Feast – Sara made happy by your letter ... allowed me to cook dinner P[arty] for some newly arrived USA anatomists, to plan sherry P[arty] for student life and to enjoy Sunday lunch with [undergraduate and postgraduates students] Bill [Humphrey], Nigel [Weiss, future Clare Fellow], Jim Massingale [actually Massengale], Paul B [Broda?], Sarah Day.*
> [Amélie to John, 23 October 1961]

197

Occasionally, there was the opportunity to lobby. At dinner,

> *Eric [Ashby, master of Clare] was probing me about Murray [Last]! Not that I could supply anything for he already had labelled him 'intelligent and neurotic' – I think there was a reference in mind, Ibadan.*
> [Amélie to John, 12 February 1961]

John confirms, 'Murray got his African job – I shouldn't wonder if Ashby pulled a wee string'.

Our contemporaries were happy to assume they could propose themselves. John's school friend, Norman Martin, is more candid than most.

> *For a long time, I have been wanting to take Valerie to Cambridge and wondered if we might come and stay with you the weekend of May 6? I have thought of coming up Friday night and leaving around Sunday lunchtime. If this should be inconvenient, I don't mind naturally and we will stay in a B and B somewhere, or a cheap hotel – anywhere you can recommend? Apart from the economic advantages of sponging on you. I would <u>really</u> like to see you all before I leave.*
> [1 April 1961]

The "cheap" would not have offended Amélie.

Patrick Gowers, after Clare with John and Robert, became a Boyd paying guest. He was neither the first nor the last in the long line of PGs but was especially liked.

> *Patrick is coming! I am highly delighted – the nicest thing of my week. Mrs Gowers wrote to ask if we could put him up until he found rooms ... Oh – nearly forgot dear Patrick. He has come. He has 6 instruments, a typewriter, a tiny suitcase of clothes and a patent German gas-lighter. It's <u>lovely</u> to see him. He is doing a PhD on Millaud.*
> [Amélie to John at Yale, 10 and 15 April 1961]

> *He plays croquet, tennis, the 'cello, the piano, the trombone and the fool with equal skill and amusement to us all ... There was a phone message y'day summoning him to London to make Music for some 'show'.*
> [Rick to John, 17 April 1961]

Patrick is staying with us at the moment, fitting gallantly into the Boyd emotions and humours. He has played midnight croquet after washing up and then accompanied while I nervously lost my way through several Beethoven 'cello sonatas. At the moment he is writing music for a negro play which is going to come on at the Royal Court Theatre (as he is paid per bar, he can best put it in 1/1 time).
[Stephen to John, 17 April 1961]

Seduced by Amélie's interest in their lives, most PGs and long-stay visitors – anxious French adolescents, young Americans finding their feet, German au pairs – were expected to become family friends, and did. Patrick was the most durable.

'Cowslips...'

Nature was central to Amélie's concept of a coherent life. Bad weather – 'it always rains in Ireland' – was balanced by good.

Terribly wet here! A field of mud. It rains like that for 12 hours but then we have the hot sun for another 12 – or so it has done to date. Took them all fishing for mackerel last evening – a wonderful evening sun, through the sea – past Portland Bill with the mackerel sizzling in the water and the boys – even Rickie (his face glowing) heaving them over into the boat.
[Amélie, camping near Weymouth with children and their friends but, as usual, without Dixon, the recipient, 17 August 1950]

The garden in the early morning, or in the evening, like whiskey "cures all but death and baffles that a long time" ... We went to Ely earlier and the Fens are dark purple and spring wheat just showing.
[to John from Cambridge, 10 April 1961]

Mummy has had mad plans for visiting the Trinity cherry trees, for the last week, at 7 o'clock in the morning. Today she got as far as the Leys before her bicycle broke down.
[Stephen to John, 17 April 1961]

I am still astonished at the calm that comes when I get into a grassy field, or hear the sea, like today.
[to John from Ballycastle (with Dixon on this occasion; on his way to an honorary QUB degree), 2 July 1961]

'No sooner thought of...'

Social vigour and promptitude were typical and could border on the hypomanic. Her, perhaps apocryphal, acceptance of three lunches on the same day followed by attending all three was emblematic. So were not infrequent demonstrations of excessive gratitude. She could extend thanks, or do-gooding, to eccentric levels.

> *Steve left on Monday but managed to fox driving examiner sufficiently to pass test and "L" plates now safely tucked away for Rick. I was so pleased, as [is] my won't [sic], with Steve's successes, that heard myself inviting his instructor home for a drink and filled him with gin!*
> [To John, 24 September 1961]

On another occasion, after a winter visit to Robert's north-west home, she took the Manchester-Ipswich coach to Cambridge (a typical travel-economy). Snow and ice led the driver to abandon the drive on reaching Cambridge, she, by then, the only passenger. The vehicle ended parked in Newton Road and he in Amélie's spare bed. At the time, such forceful eccentricities were embarrassing, especially if we were made to participate. Now we can rather admire her chutzpah.

'I shall take Bellow...'

Amélie was a keen reader of biography and novels and shared her likes.

> *Do get hold of Simone de Beauvoir's novel [The Mandarins] if you haven't read it. I think that you would find as I do that it has the wide canvas and sound portrayal of a Tolstoi novel. I am impressed and have just bought "le Deuxieme Sex!"*
> [10 April 1961]

Though not as voracious a reader as Dixon, she was attentive to both political and intellectual currents.

> *Invitation to farewell party for our Hungarian Ede [hosted as refugee from Hungary after the 1956 Soviet invasion] and his wife Pat before*

they sail for Australia. He has a good job there and so that story ends happily. Do you remember his arrival in Grange Road? How the world wags.
[10 April 1961]

Rob has given me a subscription to Encounter and in it I have just read Cyril Connolly – he seems well suited to my 54' birthday sentiments – "I believe in..."
[8 May 1961]

We are devastated over Daag [Death of UN Secretary Hammarskjold killed in plane crash] ... What next John? Give me a line on what is happening from your angle.
[24 September 1961]

Last week Crick found the pattern for nucleic acids (D.N.A.) in chromosomes which indicate genetic pattern. So now they can work on and get the faults of congenital malformations. When discoveries are published. I feel as JMcC who wanted just another ration of living to see what would be discovered next!
[2 January 1962]

'Fritter all, à la Pa...'

Expressed disappointment in Dixon, as in this letter, is vanishingly rare. We did observe a developing irritability towards him in their later years, mainly the not-unexpected response of any wife to the rubs and restrictions of life with an increasingly invalid husband (the smoker's breathlessness which increasingly impaired his well-being first became apparent before he turned 50). There seems to have been another element. A gradual fading of her student expectation that he would change the world, at least of anatomy, may have played a part. The peak of expressed negativity almost never rises above 'he is busy "stopping smoking" again'.

Worry about the children, especially Stephen and, less pervasively, Robert is commonplace. Written worries about Richard and John are rare though John occasionally needed reassurance. 'You can feel with people and all their goings on – that's what makes you able to act, to mimic, to play music and to write ... you <u>are</u> an All Rounder'. She was a persistent nagger-driver of us all. 'I'm glad you enjoyed the holiday journal that Mummy produced. She kicked us so hard to make us write something', reports Stephen. She reminds John, not for the first time that 'Rick would value a <u>letter</u>'.

Reading all this we perceive ever more clearly the privilege of a childhood overseen by this (as her grandchildren might have it) talented, semi-unemployed doctor and professor's wife. But – a big but to us – life close to Amélie was not restful. Perhaps Dixon came to feel the same with some retreat from the joy of the early letters to his department, his study and his books.

A challenge for us was the vigorous interest she took in our evolving careers, often mildly embarrassing but sometimes more. When John became a budding diplomat hoping for a post in China she, *naturally,* joined the British China Friendship Association. It had a fellow traveller attitude with close links to the Communist Party. Maybe she didn't know that or, indeed, rather approved. John's response was uncharacteristically blunt.

> *Some trouble re your flirtation with the British China Friendship Association. Too late to cry over spilt milk, and I hope no real harm done but suggest a. that you resign <u>pronto</u> from that group. Quote Helen "tell your mother not to be such an ass!" Other suggestion, that you use a bit of discretion and do not bandy my status or activities around with the Woosters, Mrs Broda, or similar cronies. I would hate to have my name become a byword with them, also the idea that they might tell all their friends to ring me up in Peking (if I ever get there and am not considered too risky) ... I am, of course exaggerating a bit, but the thought strikes me that there are 24 other new [Foreign Office] entrants who, all things being equal, are more likely to be sent to interesting posts than people whose mothers join weird groups. And worse, if you don't promise to be good, I won't even be able to tell you the unclassified interesting chat about China. Then what use w'd my letters be?*
> *Two further points*
> *a. If by any chance they send people to ask you questions – and I think this is pure fantasy; I have no real belief they would – then for goodness' sake don't bluff them. Tell them exactly what the (God knows!) tenuous enough relation with the Woosters is*

b. <u>please</u> do not, as a result of this note, start writing letters to the gov't clearing my good name (as you did when the Jade was confiscated) I can look after this one myself, and would think any initiative from you <u>very</u> unfortunate.

Finally, I love you just as much as before etc and I don't think you have really any cause for alarm, or worse, reproaches. But do, as a general rule, make sure what you're signing before you get people handing you party cards!!!

Lovest. J

P. S. How does one convince people that one's Ma is a most lovable nutcase rather than a serpent?!!

[John, probably in Hong Kong to Amélie marked "private and personal", 11 September 1962]

Those were the Woosters who had given sanctuary to John from the doodlebug and Hilda Broda, a Cambridge child health doctor who had fixed for Robert to work in her former husband's lab in Vienna [he probably had been a Soviet spy]. Nora Wooster had called Amélie her dearest sister. So far as we know, "they" did not "send people" to visit Amélie.

'Having 'let fly' I feel quite calm again...'

Only rarely did she reflect on her own feelings but, we, and later our wives, were conscious that Amélie had an anxious side, sometimes darker than anxious. One has to read between the lines for occasional hints.

> *You need not worry about your Ma who is most Sorbo [rubber]-like in her rebound and sufficiently Ulster bred to have a limited sensibility plus great quantities of toughness! ... You make me feel satisfactorily daft as opposed to daft daft.*
> [to John]

A hotelier once christened Amélie, "*Donna Strordinaria*", the title stuck. A mother *strordinaria* was a mixed blessing. As children we adored it all: the Halloween parties with ducking for apples and unusual games; the midnight swims; the bicycle holidays and the camping; our inclusion in the warmth of her many friendships. We adored even more the force and the warmth of the special love she gave us; the lift of the heart when she arrived. After adolescence adoration faded, probably from all four of us. The validating welcome of the warm white bath towel was no longer cosy but constraining. As Richard summed it up, we no longer wanted to get too close to the sun. Instead of shared delights, we sensed an aura of anxiety.

Once a widow, Amélie was a devoted grandmother but one with disappointment that the intimacy with her sons had gone. Perhaps there was an extra element. She had, she said, adored Elsa. But then, Elsa had disappeared from her life. Possibly we were repeating the pattern. Should we feel guilt?

ii   Dixon and his Ashtray

'James Dixon smokes no more
Hear the thunder in his roar
See his family scared and meek
During his no smoking week.'

Dixon always smoked. His long elegant fingers with rather flat nails which Stephen has inherited, were, on his right hand, yellow-brown. We all knew his brand was Players. Amélie's favourites were Philip Morris but she was rarely seen to light one. Dixon's periods of abstinence were short and rare; a family joke. In reality unfunny, the addiction, though we didn't call it an addiction, was deadly. It was a slow decline. Into his 40s he still bowled to us at cricket; his lankiness characteristic in his run up to the pitch. Later, we never remember him running and, on his bike, as long as that continued, it was always a slow pedal.

There were hospital assessments. Rosenheim, professor of medical student, Robert,

sent for me and I had the full colleague's treatment. X-rays and pros and cons of various treatments. Kind of him but I don't feel like assuming that level of adulthood – will have to soon I fear.
[Robert to John, probably 1961]

Four years later, Robert writes again.

> *Daddy is deteriorating fairly steadily and has become extremely hypochondriacal. Spends all the time looking at his hands to see if they are the right colour ... with any luck he will not deteriorate much over the summer; but will then get ill fairly quickly when the weather gets bad. As I see it there are three paths (1) to come home on holiday in July for a month, (2) to scrub the F.O. and come home permanently, (3) to come home as quickly as poss. after the exam at the beginning of November.*
>
> *I am against (1) because leave taking at the end of the holiday would be difficult. Against (2) because on the whole I think it a mistake to chuck one's career even for an ill parent.*
> [Robert to John in Hong Kong on Foreign Office language course, probably 1965]

Robert was over-hasty; Dixon lived another three years and neither symptoms nor hypochondria made him abandon science. In his last year he wrote seven papers, two with Richard. William came loyally and frequently to work on "The Book". Miss Salisbury bicycled every morning to take dictation or to fetch recorded tapes for transcription. Dr Anderson, the NHS GP, came daily at bedtime to ease the night with an aminophylline injection. Oxygen was provided. Sons visited at weekends.

Before his end, Dixon had sent a, somewhat characteristic, note. It is undated.

> *Following points may be of use in dealing with things and in helping and advising your mother.*
>
> *1.   Cremation. I would prefer family not to go to crematorium. Ashes don't matter but could go in Clare Garden or in our garden or in Cam (off Clare Bridge). Last has certain appeal but seems a sinful waste of phosphorus and potassium!*
>
> *2.   No service. Please tell Sir Eric [Ashby], and explain to Charles Moule [reverend professor] that I have so wished, these past several years, that I could believe in the Christian revelation.*
>
> *3.   Contents of drawers etc in lab are mine, as are also the contents of files in my room and of files numbered 1 to 120 in the library. The latter files are also my property. I think these should be claimed and sent to Dawson's, or some other book agent, for sale.*

I can state that I have given a very large number of books to the Anatomy library, and most of the reprints in my files are duplicates of material in the library.

Despite that second paragraph we had a memorial in Clare Chapel but, in keeping with that request, Richard stayed away.

Amélie remained solo in Newton Road another 20 years; a Cambridge figure, shades of widowed Raverat. She had built a bungalow in the Newton Road garden and divided the family house into flats as a source of income. She made – new style "paying guests" – friends of her tenants. They, Joan and Peggy, Dixon's former colleagues and a portfolio of those to whom she gave moral support meant she was not overtly lonely. Children and increasingly numerous grandchildren came to stay.

But, when she came to leave the home she and Dixon had shared, she howled. Child-carer Norah, now voluntary grandmother-carer, thought Amélie had never really mourned his death. Perhaps that was so.

9 Cycle of Life

As we write, it is just 100 years since Dixon and Amélie first met in 'Hughie Graham's chemistry lectures'. The Boyd Boys have become Boyd adults and then Boyd old men. John has died. Assisted by the letters the years lose their distance but there has been more than time enough for each of us to lead a life, and for us in our turn to reproduce. Might Dixon and Amélie take pleasure, or even have some pride were they able to contemplate the near-50 descendants of their union? We hope so. We are also struck by how intergenerational continuity has played out in our family as in other Cambridge families (and how persistently the advantages of our childhood continues to propagate advantage).

*17. **Boyd Boys 2** – from left: Stephen, John, Richard, Robert; party for Robert's 80ᵗʰ birthday, 2018.*

Richard became a physiologist, Oxford medical don and Brasenose fellow. His teaching was of medical students. Research into the biological transport of nutrient molecules included study of the placenta, Dixon's favourite organ.

Stephen, when still at Bedales, told John he 'would like to learn Chinese or Arabic'. In the event it was to Japanese that he gave his allegiance, settling there as bilingual lexicographer and professor of English language at Handai (Osaka University). Many whom he taught were medical students.

Robert became a children's doctor and paediatric professor. He researched placental function, and became a medical school and NHS leader. Occupations, he noted, initially with embarrassment, later with wry amusement, which copied and merged the activities of both parents.

John – "Gravitas Boyd" at prep school – became a diplomat. 'A Jew in the Foreign Office,' commented Egon; the Iklé name was prudently dropped from *Who's Who*. After postings across three continents and time as ambassador to Japan, becoming master of Churchill College and honorary fellow of Clare circled him back to Cambridge. A period as chair of the British Museum Trustees also seemed a not inappropriate nod to Iklé "art and culture". His similar role for the Needham Institute, named for the Needham who had led the British China Friendship Association which worried embryonic diplomat John raises a retrospective smile. If, in his absence, that career sounds stiff or pompous, John was not. He was a warm, witty, generous and lovely brother.

At John's 80th birthday we reflected on a letter of Dixon's, written 30th June 1944 at the height of the doodlebugs but happily un-delivered. Let this letter have the last word.

Dear John, Robert and Stephen,

The three [only three then. Richard was conceived two months later, the week Talbot Road was hit] of you have given your mother and me very great pleasure and I hope, more than anything, that she and I will live long enough to see all of you grow up and settle down in manhood with good starts to what will be happy and successful lives. But these are dangerous times and bad luck might suddenly leave you without one or other or both of us. That's the reason for this letter written while robot bombs are passing overhead and exploding (some near but mostly in the distance!)

Be good brothers to each other and remember always that you share between you the closest of human relationships – a common heredity. There is nothing more extraordinary in this world than the mingling of the characters of two individuals who have derived those characters from countless ancestors, and the production of a new generation. The three of you are the living testimony to the love your mother and I have for each other and while one of you, or any of your descendants live neither she nor I will be completely dead. So let nothing interfere with your mutual affection and always try to help each other.

If you can, be good students, get a good education, choose your life's work carefully and work hard at it. I believe you will all have ability – that's probably paternal prejudice – and I hope you will exploit your ability and give and gain from the world those things which trained intelligence can give and gain.

Choose your wives carefully and don't be taken in by superficial charm. The best thing that can happen to a man is the choice of the right wife.

Don't accept traditional or current opinion just because it is traditional or current. Use your minds and don't be hypocrites. Be as true to yourselves as you can, self-deception is very easy and quite soul destroying.

Remember always that you have given us, your mother and me, great pleasure – and that we are proud to have been your parents. I can only hope you need never see this letter but if you do I wish the three of you happy lives and joy in living.

Index of Names and Information on Individuals Mentioned

The names of the authors and their parents are not included. Individuals who were children at the time described are usually entered under their parents' names.

Adams, Sir Walter 1906-75, Historian. Sec. Academic Assistance Council. University Admin. Director LSE — *187, 191*

Adrian, see Keynes

Adrian (Lord), E D 1889-1977, FRS Physiologist. Nobel 1932. Master of Trinity College. Husband of Hester — *82-3, 158, 161, 165, 169-70*

Adrian, Hester 1899-1966, Mother of Anne (Keynes) and Richard +1 — *169-70*

Adrian Richard FRS 1927-1995, Physiologist. Master of Pembroke College — *165*

Aileen, see Sproule

Allworthy, 1866–1952, Belfast GP/dermatologist/radiotherapist. Stepfather of Eddie Bennet — *37*

Amis, Sir Kingsley 1922-95, novelist, *Lucky Jim* 1954, +20. Swansea and Cambridge — *12, 132, 197*

Ansell, Norah, Nanny/child-carer London and Cambridge — *83, 135, 139-143, 154, 189, 206*

Ashby, Eric (Lord Ashby) FRS 1904-92, Botanist, VC Queen's Belfast, then Master Clare College — *190, 198, 205*

Auntie Amélie, see Lewandowsky

Aunts, see Day, Fink, Lowenthal,

Baker, John (Later Lord Baker) FRS 1901-85, Cambridge prof. and Clare fellow, Structural engineer — *78, 190*

Barcroft, Bridget (Biddy) née Ramsey 1906-1990, 'clinic' doctor. Wife of Henry, ch: John, Michael +2 — *42, 55, 64-5, 103, 112-3, 122, 134, 138-9, 167*

Barcroft, Henry FRS 1904-98, Professor of physiology Belfast then London (St Thomas's) — *55, 65, 112, 163, 185, 193*

Barcroft, Sir Joseph 1872-1947, FRS Professor of Physiology Cambridge 1925-37. Father of Henry. *109, 131, 164, 172, 179*

Barcroft, Mary (*Mill*) née Ball. Wife of Joseph *112-3*

Barron, D H (Don) 1905-93, reproductive biologist, Cambridge then Yale. *62, 67*

Bartley, James (JOB) 1906-67, Professor of English Calcutta, Senior lecturer Swansea *12, 29, 75, 132, 147, 197*

Baxter J S 1901-60, Walmsley anatomy student. Anatomist in Cambridge, then Prof in Cardiff *50*

Bellairs, Angus 1918-1990, professor vertebrate morph. St Mary's London, herpetologist. Wife, Ruth *58, 69-75, 94, 107, 115*

Bellairs, Ruth née Morgan FRS 1926-2021, professor University College London, chick embryologist, *73*

Bennet, Edward (Eddie) MC 1888-1977, Army chaplain, Jungian psychiatrist. Distant cousin Amélie *24, 26, 35-6, 42, 44, 71, 88, 91, 126*

Bennet, Glin 1928-2015, surgeon then psychiatrist. Son of Eddie *91*

Bennett, Alan 1934- Oxford graduate and historian. Playwright, author and actor. *168*

Bergson, H-L 1859-1941, French analytical philosopher *15*

Bertram, May 1905-1981, Medical missionary in Sudan, aunt of Roger – see Rick's Gang. *169*

Bethe, Hans 1906-2005, Nuclear Physicist Cornell, stellar energy Nobel 1967 Wife, Rose née Ewald *161-4*

Biddy, see Barcroft

Blunt, Sir, (knighthood withdrawn and FBA, resigned) Anthony 1907-83, Art historian, Soviet spy *90*

Boase, T S R (Tommy) FBA 1898-1974, Art Historian *89*

Bodian, David 1910-92, American scientist; Bodian silver staining of neurons, polio research *59*

Bodmer, Sir Walter FRS 1936- German-born Oxford professor. Polymath, geneticist, Public understanding *180*

Bohr, Neils 1885-1962 Danish nuclear physicist *163-4*
(Nobel 1922)

Bomford, W (Bruce) 1919-2004, Surgeon, RAMC *184*
D-Day, Chief medical officer BP

Bonn, M J 1873-1965, Liberal German economist *15*

Boothroyd, Betty 1929-2023 former dancer, Labour *111-2*
MP then Speaker

Bourne, G H 1909-88, Australian-born anatomist *128*
and primatologist. London Hospital (with Dixon),
Emory

Boyd, Annie 1879-? Dixon's aunt. Mother of *29*
Norman and Gerald Townsley

Boyd, Grace née Smythe 1880-1914, Dixon's *1, 29*
mother

Boyd, James Dixon 1881-1945, Dixon's father m *xi, 1, 28-9*
1.Grace; 2.Evelyn; 3.Lilly

Boyd, Lillian (Lilly) née Gledhill 1890-1964, *1, 5, 155*
Dixon's second stepmother, sons *Kenny, Lindon*

Bragg, Sir (William) Lawrence MC CH FRS *10, 162*
1890-1971, Cambridge, X-ray crystal diffraction
Nobel 1915

Bragg, Sir William OM PRS 1862-1942, Physicist *162*
Leeds, London. Nobel (with son) 1915

Brandon, H P of Brandon and co. Belfast, Dixon's *67*
accountant (and probably of JMcC)

Brignell, Emily (Nanny). East Cambridge *51-5, 69, 75, 77-81,*
child carer, retired with Folkestone cousin - *100, 108, 135-7,*
'Mrs Bachelor' *140-4, 189*

Broda, Ernst 1910-1983, Austrian physical chemist, *197, 203*
War-time in Cambridge. Possible Soviet agent. Son
Paul +1

Broda, Hilde (or Hilda) 1911-, School Medical *202-3*
Service Cambridge. Husbands
1. Ernst, 2. Nunn May.

Bronowski, Jacob FRS 1908-74, FRS *181*
Mathematician. Statistician. Operation research TV
science programmes

Brooke, Katie née Symmers. Childhood and lifelong *68*
friend of Amélie

Brooke, Rupert 1887-1915, Cambridge Apostle, *xii, 163, 170*
Fellow of Kings College, poet

Brunschvicg, Leon 1869-1944, professor of *15*
Philosophy, Sorbonne (1909-40), dismissed as
a Jew

Bullard, Sir Edward (Teddy) FRS 1907-80, nuclear *189*
then geophysicist (marine). Fellow Clare, Churchill

Butters, DG Burma Star 1915-xx., Headmaster *123, 126, 133*
Kings Choir School (1950-58)

Calvert, Raymond 1906-59, QUB student. Writer of *12-3, 147*
The Ballad of William Bloat

Camps, Francis 1905-1972, Professor of Forensic *154*
Pathology London Hospital Medical College

Cantelo, April 1928-2024, Opera singer, soprano. *171*
Married to Colin Davis 1949-64

Carrell, Alexis 1873-1944, French, innovator in *33-4*
vascular surgery and tissue culture, Nobel 1912

Catherine, see Stevenson

Chamberlain, Neville 1869-1940, Prime Minister *60*
1937-40

Chibnall, Albert FRS 1894-1988, Clare Fellow, Prof *191*
Biochemistry (Supervisor Sanger), local historian

Churchill, Clementine, 1885-1977, wife of W.S. *2*
Churchill (She was not Lady Churchill in 1926)

Churchill, W S 1874-1965, Politician, (1926, *2, 60, 67, 81*
awarded an Honorary degree as QUB Chancellor)

Clark-Kennedy 1893-1985, Physician London *191*
Hospital, Dean London Hosp. Medical College
1937-53

Coates, Stephen Child Psychologist, Hospital for *121-2*
Sick Children. Research on hydrocephalus

Cockroft, Sir John 1897-1967 (Nobel 1951) *190*

Cohnheim, JF 1839-1884, Pathologist. Pioneered *148*
role of white cells in inflammation

Connolly, Cyril 1993-1974, Literary critic *201*

Cooke (or perhaps Cook), Phyllis, of 'Miss Cooke's' *54, 68, 75, 79, 100*
Cranmer Road nursery school; daughter of A B
Cooke

Cooke, Arthur B 1868-1952, Professor of classical   *54*
archaeology 1931-4

Cook, Peter 1937-1995, Comedian; Pembroke   *168*
College graduate

Cosslett, VE FRS 1908-90 'Founder of Cavendish   *191-2*
Electron Microscopy Department'

Crawford, Sir Theo 1911-93 Pathologist   *154*

Crichton-Miller, H 1877-1959. Psychiatrist, founder   *37*
of Tavistock Clinic (originally for shellshock)

Crick, Francis OM FRS 1916-2004 Cavendish   *162, 191, 193, 201*
Cambridge, DNA (later Neurosci) Nobel 1962
Ch. Michael +2

Dale, Sir Henry OM PRS 1875-1968 London   *162*
(NIMR) Pharmacologist, nerve transmission
Nobel 1936

Darwin, Charles 1809-82, naturalist, *Origin of*   *xii, 170, 174, 176*
Species

Davies, E. Beresford, 1913-2001, Clare graduate   *196*
(with 'dining rights'), Cambridge psychiatrist,
formerly RAF

Davis, Sir Colin 1927-2013, Conductor. Married   *171*
1 April Cantelo. 2. Shamsi

Day, Joan née Loewenthal 1911-2004, artist.   *2, 18, 21, 28, 31, 61,*
Sister of Amélie (Children Sarah and Thomas)   *65, 86-7, 93-5, 139,*
    197, 206

Day, T D (Tom) 1907-75, Pathologist Cambridge   *65, 73, 94-5*
then Leeds. Husband of Joan

Deakin, Sir William DSO 1913-2005, Hist., Warden   *79, 100, 122*
St Anthony's. Ch. (by Mgt, later Hodson) *Nick,*
Mikey

Dean, H R *(Daddy Dean)* 1879-1961, Professor   *10, 182*
of Pathology (appointed before introduction of
retirement age)

de Freitas, Sir Geoffrey MP 1913-1982, Children   *93, 101, 111-2, 127,*
Frankie and *Gra* (Attlee godson) +2   *131*

de Freitas, Helen née Bell 1910-1998, US born wife   *111-2, 195*
of Geoffery.

Dempsey, Ed 1911-75 US anatomist and   *192*
placentologist

Dimbleby, Richard 1913-65, BBC
war-correspondent   *149*

Doggart, John 1911- xx, QUB student, later CEO   *13, 146*
Friedland – chimes, dolls and footballs.

Doggart, Sara (née Friedland) 1914-2015,   *13, 197*
Russian/British, Sasha dolls, photographer.
Husb. John, Ch. *JV +2*

Dorothea, see McDowell

Ehrlich, Paul 1854-1915, Immunologist; inventor of *148*
chemotherapy. Nobel 1908

Elliot Smith, Grafton 1871-1937 anatomist and   *176*
anthropologist

Ewald, Clara (*Dotta*) 1859-1948, Portraitist (Rupert *163*
Brooke, Albert Schweizer, Loewenthals!) Mother of
Paul

Ewald, PP (Paul) FRS 1888-1985, Stuttgart, Belfast, *133, 163*
Princeton. Crystallographer Wife Ella

Ewald, Ella née Phillipsohn, Wife of Paul. Sister   *162-3*
of Hansi. Mother of Rose (Bethe) and others

Fink, Egon 1906-71, husband of Peggy. worked   *92-3, 95, 208*
for American Joint Distribution Committee, 'Joint'

Fink, Margaret (Peggy) née Loewenthal)   *2, 25, 65, 68, 77-8,*
1913-2013, sister of Amélie (Daughter *Michelle*)   *86-7, 92-5, 140-1,*
                                                    149-50, 206

Fitzgerald, F Scott 1896-1940, American author,   *43*
Tender is the Night

Florey, Sir Howard OM PRS 1898-1968,   *162*
Physiologist/Pathologist Sheffield, Oxford.
Penicillin, Nobel 1945

Foot, Sir Hugh (Lord Caradon) 1907-90, Colonial   *188, 191*
administrator, UN rep. (Brother of Labour Party
Leader)

Forbes, Max 1906-xx, District Officer Uganda then   *114*
Librarian Scott Polar Research Institute

Forbes, Evelyn née Ferrar. Field Geologist Africa. *114, 197*
Daughter of polar explorer, Wife of Max, father of
Nick +2

Forel, Auguste. 1848-1931, Swiss neuroanatomist   *43*
and psychiatrist. Eugenicist (now controversial)

Forel, Oscar 1891-1982, founder of Prangins Psychiatric Hospital. Son of Auguste   *43*

Forster, E M (*Morgan*)1879-1970, friend of HOM, novelist and critic. *Passage to India*   *10, 164, 197*

Fox, HM (Monro) FRS 1889-1967, Zoologist Birmingham, London (Cambridge). Editor *Biological Reviews*   *158, 160-1*

Fozzard, JF Hon FRPS 1905-93 Photographer Anat., au. *Professors of Anatomy in the University of Cambridge*   *129*

Frank, Anne 1929-45, teenage diarist, died Belsen   *150*

Frankie, see de Freitas

Freud, Anna (Miss Freud) 1895-1982, Child Psychoanalyst, daughter of Sigmund   *121-2*

Fritz see Iklé

Gairdner, Sir William 1824-1907, Professor of Medicine Glasgow   *29*

Gale, John c 1943-, Bedalian pianist.   *115*

Galton, Sir Francis FRS 1822-1911, Polymath: statistics, eugenics. Cousin of Charles Darwin   *149*

Geoff, see Wooster

Godwin, Harry FRS 1901-1985, Clare fellow, botanist, father of quaternary ecology, (and of David)   *53, 55-6, 58, 125, 129, 131, 190*

Godwin, Margaret née Daniels 1898-1989, paleo-botanist. Collaborator with, later husband, Harry   *55, 63-4, 67*

Goebbels. Joseph 1897-1945, Nazi Minister of Propaganda   *149*

Goering, Hermann 1893-1946, WW1 Air Ace. Head of *Luftwaffe*, Leading Nazi   *149*

Goldby, Frank 1903-1997, Anatomist, Adelaide and London (St Mary's)   *73, 75*

Gowers, Patrick 1936-2014, Clare undergrad, English composer. Husband of Caroline, Father of Timothy +2   *196, 198-9*

Gowers, Sir William (Timothy) FRS 1963- , mathematician Cambridge professor. Fields Medal 1998   *133*

Graham, WF (Billy) 1918-2018, American
evangelical preacher. Led eight-day mission
to Cambridge 1955   *164*

Graham Hugh[ie], Chemist QUB (no further   *1, 6, 207*
information found)

Gray Sir James MC FRS 1891-1975, Prof. Zoology   *161*
Cambridge 1937-54. President Eugenics Soc.
1962-65

Greaves, Ronald 1908-1990, Cambridge Professor   *49, 63, 189*
of Pathology 1962-75

Grillo, Adesanya 1927-98, PhD supervisor Dixon.   *117, 152, 186*
Prof Ibadan, Ife (Foundation Dean), Fellow of
Churchill.

Hamilton, William J 1903-75, Professor of Anatomy   *13, 29, 84, 96-8, 141,*
Glasgow and London (Dean Charing Cross)   *176, 180, 187-8, 193,*
                                                           205

Hammerskjold, Dag 1905-1961, UN Secretary   *201*
General, died in mysterious aeroplane crash

Hammond, NGL (*Nick*) DSO CBE FBA   *49, 190*
1907-2001, Clare Fellow, Classicist, SOE
in Greece, Headmaster

Hardy, Olga née Loewenthal 1870-1955, sister   *26, 94*
of JMcC. (With her sister Elsa referred to as
The Aunts)

Harper, W F Anatomist London Hospital Medical   *187*
College then Prof. University College of the West
Indies

Harris, Geoffrey FRS 1913-71, Anatomist   *38, 183*
Cambridge, London, then Oxford, 'father of
neuroendocrinology'

Harris H.A. 1886-1968, Cambridge Professor of   *xv, 32-3, 50, 62, 81,*
Anatomy 1934-51   *98, 183*

Harrison, L. J. Bursar, Clare College   *58*

Harrison, Sir Richard FRS 1920-99, Professor of   *179*
Anatomy London then Cambridge 1968-84

Heard, Dorothy 1915-2014, Psychiatrist, Director of   *85*
Medical Studies Girton College

Hedy/Hedi see Kerpen

Heffer, Rueben 1908-1985, Cambridge publisher and bookseller — 108

Henderson, Sir David, 1884-1965, Edinburgh psychiatrist. Co-author *Textbook of Psychiatry* — 45

Hicks Sir John 1904-89, Economist Cambridge, Manchester, Oxford. Nobel 1972 Wife Ursula — 164, 187

Hicks, Ursula (née Webb)1896-1965, Economist Oxford co-founder *Review of Economic Studies* — 164, 187

Hill, Archibald V (AV) FRS 1886-1977, Physiologist and biophysicist UCL, Nobel 1922. Husband of Margaret — 10, 151, 165, 169, 171-2, 181

Hill JP FRS 1873-1954, Embryologist, Professor of Embryology and Histology UCL — 176

Hill, Margaret née Keynes 1885-1974, Founder, Hill Homes for the old. Wife of AV — 78, 83, 147, 165-7, 171, 195

Hill, Mark 1945- , botanist, Grandson of AV and Margaret — 106

Hines, Marion 1889-1982, American Neuroanatomist, Emory University — 4, 5, 18, 84, 86, 95, 128, 140, 161

Hodson, Donald. Director Overseas Service BBC. Second husband of Margaret — 122

Hodson, Margaret née Beatson Bell, then Deakin (*Catto*). Five children (one adopted) including *Nick* and *Mikey* — 79, 100, 122

Home, Sir Alec Douglas-Home 1903-1995, Prime Minister 1963-4 — 11

Horowitz, Vladimir 1903-1989, Russian/American pianist — 3, 43

Hoyle, Fred FRS 1915-2001, Astrophysicist Director Cambridge Institute of Astronomy — 164

Hughes, AFW (Arthur) 1908-75, Cambridge Anatomy then USA, *The Mitotic Cycle* 1952, Crick supervisor — 192

Humphrey, Gerald 1904-1995, Headmaster, The Leys School. Wife *Peggy*, US born. Son *Bill* — 124, 197

Hutton, Sibyl née Schuster – supporter of *kindertranspor*t. wife of Prof R Hutton, metallurgist, Clare fellow — 59, 102

Huxley, Frances 1884-1969, Founder Fellow   *36*
RCOG. Hon. Sec. Medical Women's Federation
1917-18

Iklé, Charles F., 1879-1963, Amélie's *'Uncle*   *3, 4, 19, 33-4, 43,*
Charlie'. New York businessman, art collector   *63-4, 88, 120, 152*

Iklé, Elsa, see Loewenthal

Iklé, Fritz 1877-1946, cousin of Elsa, St Gallen Iklé   *18, 21, 28, 60*
Freres. Children: Fritz (Fred), Judith, +2.

Iklé, Fritz (later Fred) 1924-2011 Arms control   *196*
sociologist, US Under Secretary of Defence
1981-88

Iklé, Judith, daughter of Fritz. Nurse   *60*

Isitt, David 1928-2009, Priest. Kings College   *118*
theologian

Jackson, RM FBA 1903-86, Professor of the Laws   *103*
of England, Cambridge. Father of *Sean*

Jacobson WU 1906-2000, Heidelburg refugee.   *151*
Cambridge Professor Experimental Medicine Wife
'Trudi'

Jacoby, John (*JJ*)1869-1953, Iklé Freres London,   *46, 152*
m 1. Olga Iklé (4 adopted children) 2. Edith
(ch.Tony +1)

Jacoby, Olga née Iklé 1874-1913, Amélie's   *3*
(maternal) aunt. Letters published as *Words in Pain*

Jean Marie, see Kieffer

Jebb, Eglantyne 1876-1928, Social reformer,   *166-7*
founder of *'Save the Children'*

Jefferson Sir Geoffrey FRS 1886-1961, Surgeon   *30, 38*
in Manchester, pioneer of neurosurgery

Jenks C Wilfred 1909-1973, Director-General,   *14, 17*
International Labour Organisation 1970-3

Jill, see Thistlethwaite

JMcC, see Loewenthal

Joll, Cecil 1885-1945, Surgeon. On Council Royal   *37*
College of Surgeons. 'Unrivalled' thyroid surgeon'

Jung, C G 1875-1961, Swiss psychiatrist and   *71, 92*
founder of 'Analytical Psychology'

Kanagasuntheram, Ragunathar 1919-2010, Anatomist. PhD Cambridge. Prof Singapore, Dean Jaffna — *186*

Kathleen, Irish maid, surname now unknown

Kaverne, E B (Barry) FRS 1942- , Cambridge zoologist of animal behaviour — *179-80*

Keal, Archie. Teacher at Kings Choir School in 1950s — *104*

Keir, Sir David 1895-1973, QUB Vice-chancellor 1939-49, Master of Balliol College Oxford 1949-65 — *63*

Keith, Sir Arthur FRS 1866-1955, Anatomist , anthropologist — *37*

Kember, 'Mrs'. 'Daily' at 21 Newton Road — *143*

Kennedy, James (*Mr Kennedy*) 1842-1933, Teacher, then Unitarian Minister in Larne 1878-1926 — *6, 8, 28*

Kerpen, Hedy (Hedi), Austrian refugee, Boyd child-carer and 'help', later Cambridge dental-assistant — *83, 135, 138-41, 189*

Keynes, Anne née Adrian. 1925-2017. Singer. Wife of Richard Children: Adrian +3 — *115, 170-1, 195*

Keynes, Adrian 1946-74. Son of Anne and Richard, medical doctor, school-friend of Richard — *104, 115, 129, 170*

Keynes, F A, née Brown ('Old Mrs Keynes') 1861-1958, Social activist. Cambridge Mayor 1932 — *166-8*

Keynes Sir Geoffrey 1887-1982, Pioneer of blood transfusion, Surgeon at Barts, Bibliographer — *134, 166, 169-70*

Keynes Margaret née Darwin 1890-1974, wife of Geoffrey. Children: Richard +4 — *170, 172*

Keynes, Maynard CB FBA (Lord Keynes) 1883-1946, Economist, brother of Margaret Hill — *15, 166, 168*

Keynes, Richard FRS 1919-2010, Cambridge Physiologist. Son of Geoffrey — *165, 169*

Kieffer, Jean-Marie, French student of English at Bell School and, paying guest of Boyds — *84*

King, Audrey, 'monthly nurse' who looked after newborn Richard — *79*

Kinnaird, Elizabeth Holmes 1894-19xx, Belfast artist and art teacher. Loewenthal connection — *34, 85*

Klaus, Liesel, Student contemporary of Dixon in Geneva at League of Nations *16*

Knight, Maxwell OBE 1900-68, MI5 1924-56, chief 'agent-runner'. Radio-naturalist *117-8*

Koch, Robert 1843-1910, Bacteriologist. Nobel 1905 *148*

Krebs, Sir Hans FRS 1900-81, Biochemist (Kreb's cycle). Berlin, Cambridge, Sheffield, Oxford. Nobel 1953 *129, 130*

Langdon-Brown, Sir Walter 1870-1946, Regius Professor of Physic, Cambridge. Sister: 'old' Mrs Keynes *37, 166-7*

Last, Murray 1936- , Medical anthropologist. Clare, Yale, Ibadan (PhD). Later, UCL Professor *198*

Le Gros Clark, Sir Wilfred FRS 1895-1971, Oxford Professor of Anatomy 1934-62; palaeoanthropologist *33, 56, 82, 176, 180-2, 188*

Lenway, Fred (Lewandowsky by birth), Californian businessman. Son of Georg, cousin of Amélie *151-2, 195*

Lenway Anne, née Baerwald. Social worker in Berkeley. Wife of Fred *195*

Lewandowsky (..ski), Amélie née Iklé 1882-1963, Amélie's aunt. See Waldeck *26, 150-1*

Lewandowsky (...ski), Georg 1880-1950, Hamburg Banker then USA. Husband of Amélie *150-2*

Lewisohn, Sam 1884-1951, New York lawyer, financier, philanthropist and art collector Iklé relation *4*

Lewis, Warren H 1870-1964, Embryologist, Carnegie Institute and Johns Hopkins *192*

Lewis, P R (Peter) 1924-2007, Cambridge Physiology and Anatomy Depts. Histochemist *75, 191*

Livingstone, Hon Millicent , Wife (Lady Livingstone) of Sir Richard *26*

Livingstone, Sir Richard (Dickie) 1880-1960, Classicist, QUB Vice-chancellor 1924-33, later Oxford *26*

Loeb, Jacques 1859-1924, German/American
biologist, father of Leonard and Robert — *3*

Loeb, Leonard 1891-1978, Physicist UC Berkeley,
once husband of Marion Hines — *3, 4, 95*

Loeb, Robert (*Bobbie*) 1895-1973, Professor
Medicine, Columbia — *3*

Londonderry, Edith, Marchioness of Londonderry
DBE1878-1959. Political hostess. — *2*

Loewenthal, Elsa née Iklé 1874-1959, Twin:
Olga Jacoby. Brother: Charles Iklé. Husband:
John (JMcC) — *xi, 2, 19, 24, 27-8, 42, 44-5*

Loewenthal, John McCaldin (JMcC, occasionally
Jack)1864-1951, father: Helen, Amélie, Joan,
'Peggy' numerous

Lowenthal, Helen née Loewenthal 1904-93, Art
Historian, Amélie's sister — *2, 24, 26, 28, 31, 43-4, 66, 71, 84-8, 90-2, 107, 139, 202*

Lucas, Keith FRS 1879-1916, Cambridge
Physiologist — *103*

MacConaill, Michael 1902-87, Anatomist QUB,
Sheffield, Prof. Cork. Republican 'Hero' of the
Battle of Belfast — *145-6*

McCullagh, Graham (*Spud*) 1904-1957, Pathologist:
QUB, Cambridge, Senior Tutor Queens College — *25*

McDonnell, Hon Hector 1947- , Irish painter — *13*

McDowell, Dorothea. Belfast drama critic, friend of
Helen Lowenthal and Amélie Boyd — *63-4, 125*

Margaret, 18xx-1948, much loved Lennoxvale
cook. (or see Godwin) — *47, 136, 150*

Marion, see Hines

Martin, Archer CBE FRS 1910-2002, Chemist,
Chromatography. Nobel 1952. Brother of Nora
Wooster — *163*

Martin, *Norman* Anglo-Spanish Westminster
school-friend of John — *198*

Massengale, JR (*Jim*). American musicologist,
Clare College, then professor at UCLA — *197*

Masson, Sir Irvine FRS 1887-1962, Chemist. *187*
Vice-chancellor Sheffield University 1938-53

Maynard, Mrs, Housekeeper at Fernside *60*

Medawar, Sir Peter OM, FRS 1915-87, Zoologist, *178-9*
transplant immunologist. Nobel 1960

Meredith, Ralph 1908-2000, Mathematician, *9, 12, 16, 29, 146-7*
student-friend of Dixon, son of Hugh

Meredith, Adam 1913-76, Professional bridge- *9*
player, son of Hugh

Meredith, Hugh Owen (HOM) OBE, 1878-1964, *6, 11, 13, 28, 70, 85,*
Economist, Professor at QUB, Cambridge Apostle *103, 168*

Meredith, Peggy née Ritchie, then Spring Rice *10, 11*
1896-1983, London GP, third wife of Hugh

Miller, Sir Jonathan 1934-2019, Medical doctor. *168*
Theatre and opera director, TV personality

Molotov V 1890-1986, Prime Minister, Commissar *114*
for Foreign Affairs
USSR

Moore, Dudley 1935-2002, Comedian and *168*
composer. Oxford graduate

Moore, George 1852-1933, Cambridge Philosopher, *9*
Principia Ethica

Moore, *'Minnie'*. Maths teacher at The Leys *128*

Moore, TW (Terry) Developmental Psychologist, *84, 120-1, 123*
Univ. London, Aarhus. *Educational Theory* 1974

Moore, Mary, wife of Terry. Children: Colin, Alison, *84-5, 121*
Arnrid

Mosley, Sir Oswald 1896-1980, British Labour *146*
Party minister then Leader British Union of Fascists

Mossley, at League of Nations Summer School with *14*
Dixon

Mossmann 1898-1991 US zoologist and *97*
placentologist

Moule, CFD (Charlie) CBE FBA 1908-2007 Clare *205*
Fellow, Lady Margaret Professor of Divinity

Murray, Gilbert OM 1866-1957, Australian-born *15*
Classicist. Public figure and internationalist

Nabokov, V 1899-1977, Russian/American Novelist *107*
(and Lepidopterist)

Nanny, see Brignell

Navaratnam, Visvanathan 1933- Anatomist, Fellow *186*
of Christs College, director of Medical Studies

Newman 'Mr', Technical manager at Heffers *108*
Cambridge Press

Norah see Ansell

Nunn May, Alan 1911-2003, Physicist, atomic spy. *202*
Later Professor Ghana. Wife: Hilde Broda

Ogilvie, Sir Frederick 1893-1949, VC QUB *49, 147*
1934-38, Director-General BBC 1938-42

Olga, see Jacoby, Hardy

Sir Denys Page 1908-78 Cambridge Classicist *190*

Palmes, Brian (*CP*). Commander RN 'Fought at *125*
Battle of Jutland'. White Hare ski club Andermatt

Pat, see Cameron

Paul, Margaret (née Ramsey) 1918-2002, PPE tutor *122*
Oxford. Sister of Bridget Barcroft

Pavlova (Anna) 1881-1931, Russian ballerina. *15*
Pudding named for her!

Pearce, AGE 1916-2003, Histochemist, Professor *185*
RPMS, pathologist Hammersmith Hospital

Pearson, Karl FRS 1857-1936, Pupil of Galton, *149*
biostatistician, eugenicist. 'Declined knighthood'

Penris, Nia. 'Mothers help' in London then family *142*
friend

Pevsner, Sir Nikolaus FBA 1892-1983, German *91*
born British architectural historian

Phillipsohn, Hansi. Refugee, former Gymnasium *133, 192*
Head; sister see Ewald

Pickering, Sir George FRS 1904-80, Regius *182-3*
Professor of Physic (medicine) Oxford

Pippard, Sir Alfred (Brian) FRS 1920-2008, Clare *161, 191*
Fellow, Cavendish Professor of Physics

Potts, Malcolm 1935- , Reproductive sci., PhD with *147*
Dixon. Prof Univ. California. Leader in population
control

Praeger, Rosamond (Praegie) 1867-1954, Ulster
sculptor   *28*

Pringsheim, EG 1881-1970, Botanist Berlin, Prague,   *126*
Cambridge, Göttingen. Son: Wolfgang

Pringsheim, Wolfgang, Medical doctor, Leys   *126*
school-friend of Robert

Prokofiev, Sergei 1891-1953, Russian composer   *3*

Ramanujan, Srinivasa FRS 1887-1920,   *152*
Mathematician, number theorist. First Indian
Fellow of Trinity

Ramsey, Arthur 1867-1954 Mathematician, Fellow   *112*
of Madgelene

Ramsey, Frank 1903-30. Mathematician and   *112-3, 122, 166, 169*
philosopher. Apostle, Fellow of Kings. Sister,
Biddy Barcroft

Ramsey, Lettice 1898-1985, Ramsey and Muspratt,   *113*
Cambridge photographers. Husband: Frank

Ramsey, Michael 1904-88, Pres. Cambridge Union,   *112*
Archbishop Canterbury. Sister: Biddy, brother: Frank

Raverat, Gwen 1885 – 1957, Artist and author,   *xii, 47, 49, 96, 105,*
Period Piece   *109, 111, 115, 157,*
164-7, 206

Ravel, Maurice 1875-1937, French composer   *3, 19, 197*

Reddaway, Brian FBA 1913-2004, Cambridge   *111, 161, 190*
Economist, Clare Fellow, Dir. Dept. Applied
Economics 1955-69

Reddaway, Barbara née Bennett 1911-96, Wife of   *103, 110-1, 131, 195*
Brian, Mother of *Peter, Stuart, Lawrence and Jackie*

Ribbentrop, J von 1893-1946, German Diplomat   *53*
and Nazi Foreign Minister

Richards, 'Dr.', GP Grange Road Cambridge. Father   *136*
of *Barry* and *Andrew*

Richardson, Sir Ralph 1902-83, British actor in   *24*
theatre and film

Rieu Dominic MC 1916-2008, Classicist,   *133*
headmaster

Rieu, E V 1887-1972. Classicist. Founder editor of   *133*
Penguin Classics. Children: Dominic +3

Rick's Gang 1952:   Basil Postan 1946- Banker.   *104*
Friend even if leader of his
own gang.

Graham Rook 1946-
Microbiologist

Mark Griffith 1946- Classicist
at Berkeley

Martin Hopkinson 1946-
Art historian

Roger Bertram 1946 -
Schoolteacher, then GP

Rodgers, W R (*Bertie*) 1909-1969, Presbyterian   *12, 13, 147*
minister, poet and broadcaster.

Roosevelt, FDR 1882-1945, 32nd American   *3*
President

Rothschild, NM Victor (Lord) GM, FRS 1910-90   *118*
Biologist of sperm, wartime spy, head of CPRE
1971-4

Rueff J L 1896-1978, French economist. Critic of   *15*
Keynes.

Rutherford, Ernest (Lord) OM PRS 1871-1937, N.Z.   *10, 191*
physicist, Prof Manchester then Cambridge. Nobel
1908

Russell (Lord), Bertrand FRS 1872-1970,   *164*
Mathematical philosopher

Salisbury, 'Miss', Professor's secretary Cambridge   *129-30, 157, 205*
Anatomy Department

Sanger, Fred OM FRS 1918-2013, Cambridge   *162, 191*
biochemist Sequenced protein, DNA. Nobel
1958;1980

Sarah, see Trainor or Day

Schrödinger, Erwin 1887-1961, Austria and   *163*
Germany then Dublin, quantum theory. Nobel 1933.

Sherrington, Sir Charles OM, PRS 1857-1952,   *158, 160*
Neurophysiologist, Liverpool, then Oxford.
Nobel 1932

Shillingford, J P 1914-99, British cardiologist,   *59*
Hammersmith Hospital

Shute, Charles 1917-99, Anatomist and histochemist, Later Emory Egyptologist. M. x3, ch: Stephen +3   *69, 73, 75, 123, 164, 191, 196*

Simpson, Oliver, Junior Fellow Trinity College 1950s. Top floor family at St Chad's, 48 Grange Rd.   *83*

Skidelsky, Robert (Lord) FBA 1939- , Biographer of Maynard Keynes   *166*

Slater, Eliot 1904-83, British psychiatrist, Queen's Square, Maudsley. Co-author *Clinical Psychiatry*   *45*

Sloan, Pat 1908-78, Communist, author *Soviet Democracy,* secretary British-Soviet Friendship Society   *15, 16*

Spooner, ETC (*Tenny*)1904-95, Bacteriologist, Clare Fellow, Dean London School Hyg. Trop. Med. 1960-70   *49, 148, 189-90*

Sproule, Aileen, QUB medical graduate. GP in Gilford, Co. Armagh   *41, 52, 55, 64*

Spud see McCullagh

Stallybrass, *Arthur*, Gardener at 21 Newton Road.   *143*

Stark, Johannes 1874-1957, German physicist, Nazi. Nobel 1919   *171*

Stevens, *Vera*, 'Daily', at 21 Newton Road. Cycled from Harston   *143*

Stevenson, *Catherine*, Teenage friend (Charleston dancer) - didn't sufficiently fancy John   *127*

Stoker, Sir Michael FRS 1918-2013, Virologist, Clare Fellow. President Clare Hall   *106*

Stone, Molly, Headmistress Howell's (girls) School, Denbigh   *111*

Streeter, GL 1873-1948, Embryologist, Director Carnegie Institute Washington. Dedicatee *Human Embryology*   *34*

Taylor, Harold, CBE 1907-95, NZ, Clare Fellow, Mathematician, church historian, Admin head, then V-C Keele   *63, 109, 111, 126, 166, 1909, 195*

Taylor, Joan, née Sills 1903-65, Co-author *Anglo-Saxon Architecture.* Husband: Harold, ch: *John*+3   *110-1, 131, 195*

Telfer, William (Revd Canon) MC 1886-1968, Dean   *164, 191*
of Clare, Prof. Divinity, Master of Selwyn College

Thackeray, WM 1811-63, Author, *Vanity Fair*   *130*

Thistlethwaite, Frank 1915-2003, Cambridge   *127, 195*
historian, First V-C UEA. Wife: Jane, US born.
Ch: *Jill* + 2

Thirkill, Sir Henry (Thirks) MC 1886-1971,   *50, 52, 81, 130-1,*
Physicist. Master Clare College 1939-1958   *189-91*

Thompson, Dorothy 1993-1961, American journalist *3*
married to Sinclair Lewis

Thurley, Ken, Technician Cambridge Anatomy   *129*
Department, later co-author many papers

Todd, Alex, (Lord) Todd OM PRS 1907-97,   *162*
Nucleotide chem. Manchester, Cambs. Nobel 1957.
Wife: née Dale

Tower, Sarah, Johns Hopkins neuroscientist.   *161*
Collaborator with Marion Hines

Townsley, Alice (née Dickie) 1907-88, Obstetrician. *29-31, 58, 177*
Husband: Norman. Ch: Janet 1943-, Gerald 46-

Townsley, Norman (earlier Thomas) FRCS 1905-   *13, 28-31, 38, 42, 58,*
2001, Surgeon Norwich. Wife: Alice. Cousin: Dixon *98, 126*

Townsley, Gerald (earlier Bernard) FRCS 1907-75, *13, 28-31*
'Radiotherapeutic' surgeon Kent. Cousin: Dixon,

Trainor, Sarah, 'Housemaid' at Lennoxvale,   *47, 65, 141*
(longstanding and caring). Catholic

Trevelyan, Sir George 1906-1996, Educationalist. *90*
Warden of Attingham Park

Tudor-Jones, Lecturer in Embryology University of *39*
Liverpool 1930s

Turnbull, H M FRS 1875-1955, Pathologist,   *77*
Director London Hospital Institute of Pathology
1906-46

Uncle Charlie, see Iklé, C F

Uncle Fritz, see Iklé, Fritz

Vining, Joe 1938- , Clare and Harvard. Lawyer US *197*
Justice Department then Law firm

Waldeck, Clara née Lewandowski 1914-1997,   *150*
daughter of Georg, wife of Hans

Waldeck, Hans 1899-1942? Dutch businessman. Killed in France *en route* via Portugal to USA. — *150*

Waldeyer, HWG 1836-1921, Anatomist, Berlin. Coined words 'Neuron' and 'Chromosome' — *148*

Wallace, Edgar 1875-1932, Author of detective and other fiction — *25*

Walmsley, Thomas (Tommy) FRSE 1889-1951, Professor of anatomy QUB 1919-50 — *20, 25, 32, 36, 42, 50, 70, 173-6, 180*

Wathen GA CIE, Principal Khalsa College Amritsar 1915-24, Headmaster The Hall School Hampstead 1924-55 — *132, 148*

Watson-Jones, Sir Reginald 1902-72, Paediatric surgeon Liverpool and Great Ormond Street — *40*

Wedderburn (Lord), *Bill* FBA 1927-2012, Fellow Clare, Prof Commercial Law LSE, Labour Party Peer — *132*

Weiss, Nigel FRS 1936-2020, Clare graduate then Fellow. Professor of Mathematical Astrophysics — *197*

Weiss Peter 1937-99, Fellow pupil with Robert at Byron House School and lifelong friend — *125, 127*

White, T H 1906-1964, English author, *The Once and Future King* — *70-1*

Wilkie, Wendell 1992-1944, Lawyer, Republican nominee for President 1940 — *3*

Willmer, Nevill FRS 1902-2001. Professor of Histology. Designer of Clare garden, Artist. Wife: Penny. — *117, 134, 152, 161, 164, 189, 191*

Wilson, Harold 1916-95, Prime Minister 1964-70, 1974-6 — *11*

Wilson, J T, FRS 1861-1945, Professor of Anatomy in Sydney, then Cambridge — *5, 32, 174*

Winnicott, Donald 1896-1971, at Leys School. Later Paediatrician Paddington Green Ch. Hosp., Psychoanalyst — *122*

Wislocki, George1892-1956 US anatomist and Harvard professor — *192*

Wittgenstein, Ludwig 1889-1951, Austrian/British philosopher – *Tractatus Logico-philosophicus* — *169*

Woodside, Moya (née Moira Neill) 1907-xx, Social scientist, *Patterns of Marriage,* husband: Cecil — *110*

Woolf, Virginia 1882-1941, English author — *xii, 13, 183*

Wooster, Peter 1903-84, Cambridge Crystallographer. Treas. (leftwing) World Fed Sci Workers. Wife: Nora — *53, 55, 58, 63, 84, 113-4, 163, 202*

Wooster, Nora (née Martin) 1905-2000, Crystal Structures Ltd (with Peter). Mother:Tony, Geoff, Anna — *55, 69, 77, 84, 99, 113-5, 119, 163, 202-3*

Yates, Frank FRS 1902- 1994, Statistician. Operations research. Plant growth — *181*

Yale, Ann née Brett, daughter of JMcC's solicitor. Husband: David — *195, 197*

Yale, David FBA QC 1928-2021, Historian of Law Cambridge, Fellow of Christ's College — *195, 197*

Zimmern, Sir A E 1879-1957, Political scientist, League of Nations — *15, 17*

Zuckerman, Solly (Lord) OM FRS 1904-93, Anatomist, Prof. Birmingham, UK Chief Scientific Adviser — *180-3*

Acknowledgements

We thank those whose letters to our parents contributed to their lives and from which we have quoted; also those friends and relatives who know us or knew our parents who have commented on earlier drafts, or provided further letters. Our great gratitude to you all. Special thanks to Keith Ravenscroft for continuing encouragement.

We are also most grateful to Polly Warren for her diligent and thoughtful editing of an undigested draft and to Melanie Bartle for her great support and expertise in navigating through from draft to publication and for her pleasant manner in doing so.

The authors apologise for any failure in recognition of copyright and would be grateful if any omissions in this regard could be drawn to their attention.

Related volumes

Hamilton WJ, Boyd JD, Mossman H *Human Embryology - Prenatal development of form and function* Heffer 1945 (and later)

Boyd JD and Hamilton WJ *The Human Placenta* Heffer 1970

Jacoby, Olga *Words in Pain: Letters on Life and Death* 1919. Reprinted, ed Catty, Jocelyn and Moore, Trevor. Bloxham Skyscraper 2019

Fink, Margaret G *From Belfast to Belsen and Beyond* Penfold publishing 2008

Fink, Michelle and Boyd, Robert *John McCaldin Loewenthal - letters from a Victorian Commercial Traveller* Moore and Weinberg 2022

Bennet Glin *Understandings (Father, Son and Father trilogy)* 2006 Broadcast Books

The Authors

Richard Boyd lives in Oxford. A retired Physiologist he is an emeritus Fellow of Brasenose College.
richard.boyd@dpag.ox.ac.uk

Stephen Boyd lives in Osaka where he was Professor of English Language and is a lexicographer.
sacboyd@gmail.com

Robert Boyd lives near Manchester where he is emeritus Professor of Child Health.
r.boyd@manchester.ac.uk

John Boyd (died in 2019) a diplomat was formerly Master of Churchill College and before that, UK Ambassador to Japan.

www.ingramcontent.com/pod-product-compliance
Lightning Source LLC
Chambersburg PA
CBHW030822090426
42737CB00009B/832

* 9 7 8 1 8 3 6 1 5 1 6 3 0 *